T0361793

Lean Manufacturing
in the **REAL WORLD**

JEFFREY P. WINCEL
THOMAS J. KULL, PHD

Foreword by MIKE HOSEUS

Lean Manufacturing in the REAL WORLD

CRC Press is an imprint of the
Taylor & Francis Group, an **informa** business
A PRODUCTIVITY PRESS BOOK

CRC Press
Taylor & Francis Group
6000 Broken Sound Parkway NW, Suite 300
Boca Raton, FL 33487-2742

First issued in hardback 2019

ISBN-13: 978-1-4665-5789-5 (hbk)

Library of Congress Cataloging-in-Publication Data

Wincel, Jeffrey P.
People, process, and culture : lean manufacturing around the real world / Jeffrey P. Wincel, Thomas J. Kull.
pages cm
Includes bibliographical references and index.
ISBN 978-1-4665-5789-5 (hardcover : alk. paper)
1. Lean manufacturing. 2. Corporate culture. 3. Industrial management. I. Kull, Thomas J. II. Title.

TS155.W546 2013
658--dc23 2013004141

Visit the Taylor & Francis Web site at
http://www.taylorandfrancis.com

and the CRC Press Web site at
http://www.crcpress.com

To HMW

—jpw

To my son, Adam, who I hope to inspire through my work; and to my colleagues at Arizona State University, who allow me to do so.

—tjk

Contents

Foreword

My first introduction to Lean manufacturing and the culture issue was when I helped start up Toyota's plant in Georgetown, Kentucky—Toyota's first plant in the U.S. Being the 201st person hired, I was early enough in the start-up process to witness the many discussions about culture. People worried whether the Toyota managers were going to come in and make us act like Japanese. We were concerned how our culture was going to change. Mr. Cho, who is now chairman of the board at Toyota, was leading the start-up and he calmed many of us by communicating early on that there was no intent to create the Eastern culture in Kentucky. He stated that they were going to take what was best about the Toyota culture and what was best about the American–Kentucky culture, and create a culture that was best for the organization where business was being done. He used standardized work as an example of this. Standardized work is a way to document the current "best practice" of any process that we are working with. It gives repeatability, stability, and consistency. Uniformity is very much part of the Japanese culture. But, standardization doesn't stop there. The purpose of standardization is to help us identify waste and drive innovation and improvement, and innovation is something that we "American pioneers" were known for. Just as standardized work is a blending of stability and improvement, our plant was a blending of "Toyota and American" values and culture.

While that helped calm many fears, there was still a mindset problem that I experienced firsthand. I struggled early on with some ideals of the Toyota Production System (what we now call Lean). One time, as a final assembly manager dealing with a new major model change, I was battling with defects coming off the end of the line. I would call over the group leaders and berate them, saying that "we need to get better, so go fix those defects!" However, I was lucky to have a trainer-coach who helped me look at the process instead of the results. I needed to change my mindset, to think differently. Instead, he walked me "back into the process" to where the team members were doing their standardized work and struggling, and pulling their *Andon*. "This is where you need to be 'Mike-san', help these team members solve their problems." It was part of my culture to look only at results, and I learned that was the hard way to look at the process.

For you, rather than learn the hard way, this book, *People, Process, and Culture: Lean Manufacturing in the Real World* by Jeffrey Wincel and Thomas Kull, can serve as a trainer-coach, helping you and your organization make the Lean transformation by exposing where your values, or thinking, may also have some gaps.

As I've worked more and more with companies, I'm encouraged to see that people have moved beyond the tools of Lean. Most started their journey focused on the tools—i.e., standardization, visualization, value stream mapping, etc.—but we've come a long way in maturation, with most seeing that Lean is more holistic. As Jim Womack of the Lean Enterprise Institute puts it: Lean is about purpose, people, and process. "Purpose" is about everybody being on the same page as to the vision, values, and mission. "People" is about mutual long-term prosperity that's a win-win for the people and the company. "Process" is about best utilizing the Lean tools to identify problems while engaging people at all levels to find solutions. I usually add my fourth "P" which is problem solving or PDCA (plan, do, check, act). What I appreciate about this book is that Wincel and Kull extend the conversation beyond the tools, and speak directly to each of these Ps.

Culture is about vision, mission, and values. This book identifies the Lean values that will help you coordinate and integrate those within the people in your organization and your country. When I work with companies, I'm not trying to create a Toyota culture. Often, managers from companies in different countries ask me, "How does our national culture marry with Lean?" Until now, all I was able to do was tell my stories and anecdotes. I had not been able to give objective, global data to point to and quote. With this book, I am thrilled and happy to be able to do that. This book provides clear objective data, going beyond what has been said qualitatively, and instead giving quantitative evidence. This evidence shows not only that Lean values matter, but shows where and which values specifically will matter in certain national cultures.

This is a pioneering book. It will help Lean champions throughout the world. A big challenge for Lean champions and middle managers is to influence up: Engaging that executive leadership team. Personally, I too have had different degrees of success with leadership teams. This book provides data to help address this gap. You should show it to your leaders to help raise their awareness. Try doing it in terms of PDCA, identifying gaps in the thinking and mindset that are hindering the accomplishment of your leadership's goals. Lean success is dependent on the mindset and a big gap out there is being able to show leaders that the Lean values matter. This book helps get the conversation started; gets the problem solving started. By showing that values matter, leaders will become more engaged. They may not even realize how important it is. However, once they realize it, then what should they do about it? This book also helps Lean champions pinpoint where specifically their team, their leadership, and their organizations may struggle with the Lean mindset. Unlike my early days at Toyota, you can know up front where the hidden cultural barriers are. You can use this book to say, "Ah-ha! That's why we've been having some issues here!"

Before you begin reading, recall that Lean is all about PDCA, about experimentation. Know that this book follows that same process: the scientific method. As you're reading, you're literally reading an experiment; theory is presented, hypotheses are given, data are analyzed, and conclusions are reached. This is PDCA on a global scale. So keep this in mind. As you read, Wincel and Kull present charts, offer clues, and pose questions. The intent? To help you learn more about each of the four Ps I stated above; to help you with your Lean transformation; to help you create the vision and build your problem-solving culture; and to help you engage the most people possible to be successful.

This is a different book. It applies not only to global companies with facilities in different countries, but also to local companies located around the globe. Because of its link to process, people, and culture, the results inform how to do Lean business in whatever country you happen to be in. In each country, there's a continuum of highs and lows, biases toward or against the various lean values. However, no country is a "no" or a "can't." The whole point of Lean is to be explicit about what is often implicit. Culture is (by nature) hidden. This book helps make both the potential positive and negative aspects explicit in order to help identify and solve problems—which is the whole point of Lean after all.

Michael Hoseus

Coauthor of *Toyota Culture: The Heart and Soul of the Toyota Way*

Introduction

If you "google" the words *Lean manufacturing*, the results will return more than 2.76 million entries. Clearly, what was seen as a fad of Asian car manufacturers, especially Toyota, just a few years ago has become engrained in the lexicon of American business. However, if you do the same search for *Lean manufacturing culture*, the entries drop to less than 15% of the results for Lean manufacturing.* With a total of 404,000 entries, there is still a substantial body of thought on the cultural aspects of Lean, yet the significantly fewer hits on Lean manufacturing culture are extremely telling about the American view of the Lean process. When "manufacturing" is removed from the search, *Lean culture* returns 1.75 million entries.

What are we to make from these observations? Clearly there seems to be great interest in the cultural aspects of Lean manufacturing. Yet, the preponderance of the available material, research, and application centers squarely on the use of "Lean tools." The focus on these tools mistakenly misses the underpinning of the success of the Lean manufacturing approach. Kevin L. Meyer, president and founder of Superfactory Ventures LLC and coauthor of the book *Evolving Excellence*, described it this way: "Lean is not about the tools; the tools are simply tools that help what Lean is really about: people."† Understanding what Lean is really about is that people are not just necessary but are *the* critical aspect of effective and successful implementation of a Lean environment. The criticality of people belies the relative absence of consideration of the underlying personalities and cultures from which these essential people emerge. Meyer continued: "A culture of excellence must be driven, with actions and not simply slogans and statements, from the top. A culture of excellence requires knowledgeable, experienced, and creative people to continually find better ways to create customer value."‡

* Google search results February 17, 2011.

† Kevin Meyer, The culture side of lean manufacturing. Evolving excellence. February 7, 2007. (http://www.evolvingexcellence.com/blog/2007/02/the_culture_sid.html#ixzz1EF4nQmqP), accessed February 17, 2011.

‡ Ibid.

RESEARCH BASED

Nearly every company, manager, and employee has heard or told some epic, or even mythical, story about the almost unfathomable improvements made through Lean manufacturing techniques. Like many urban legends, these stories come from a guy who knows a guy ... you get the idea. Even the reference materials that exist about the actual practitioners and leaders, such as Toyota and other companies, are one step removed from the current operations. Most are from former employees or by someone trained by those former employees. In and of themselves, these expository commentaries can be extremely beneficial, provided the reader recognizes the limited scope of the underlying research and data set from which the material is obtained. The material in this book is derived from the academic research of Drs. Thomas J. Kull, Tingting Yan, and John G. Wacker from Arizona State University, W. P. Cary School of Management; from Drs. Shaohan Cai, Carleton University (Ottawa, Ontario, Canada), Sprott School of Business; Minjoon Jun, New Mexico State University; Zhilin Yang, University of Hong Kong; from Drs. Damian Power and Danny Samson, University of Melbourne, and Tobia Schoenherr, Michigan State University, Eli Broad School of Management. The source material is derived from their research articles as published in the *Journal of Operations Management.**

MANAGERIAL-APPLICATION APPROACH

From a pragmatic viewpoint, academic research, no matter how relevant, often has little practical application to the workplace. Operational thinkers and practitioners need the transference from the theoretical to the observable to make any meaningful sense of the researchers' work. *People, Process, and Culture* is exactly that kind of transfer.

* Thomas J. Kull and J. G. Wacker, Quality management effectiveness in Asia: The influence of culture, *Journal of Operations Management* 28 (3),223–239 (2010); Shaohan Cai, Minjoon Jun, and Zhilin Yang, Implementing supply chain information integration in China: The role of institutional forces and trust, *Journal of Operations Management* 28 (3), 257–268 (2010); Damien Power, T. Schoenherr, and D. Samson, The cultural characteristic of individualism/collectivism: A comparative study of implications for investment in operations between emerging Asian and industrialized Western Countries, *Journal of Operations Management* 28 (3), 206–222 (2010).

The real world of Lean processes is examined in the context of the researchers' work, observations, and data. Actionable opportunities and conclusions are drawn from that work—specifically in the context of *culture*, as an abundance of material on Lean process floods the operations world. Without an understanding of and appreciation for the culture within a company, within a country, and within a global region, Lean manufacturing systems fall on the rubble of other failed initiatives.

Lean manufacturing begins with a transformative culture, a Lean culture. Companies, and their executives and managers, must lead the cultural revolution, not in some anemic and benign way, but in a transformative and sustainable way. Nowhere is the folksy adage "lead, follow, or get out of the way" more relevant. Unfortunately, too many of those who should be following or getting out of the way believe themselves to be leaders.

Acknowledgments

A great debt of gratitude must be paid to the Department of Supply Chain Management from the W.P. Cary School of Business at Arizona State University, especially Dr. Thomas J. Kull and Dr. William A. Verdini. Dr. Verdini encouraged my continued efforts in the synthesis of academics and business management. Dr. Kull has been a great collaborator, providing the research basis for this book.

—J. P. Wincel

I am deeply indebted to Jeff Wincel for his passion for research that is relevant and available to managers and students. Without his leadership, this book would not have happened. Also, what this book presents would not exist without the efforts of Dr. Tingting Yan and Dr. Jack Wacker; both are great collaborators and friends. Much of this book is based on research from the numerous authors that we credit herein and to whom we are extremely grateful. Last, I must thank the many colleagues from my days in industry who have greatly contributed to my thinking concerning Lean. Those experiences are a never-ending resource for my research and teaching.

—T. J. Kull

About the Authors

Jeffrey P. Wincel is an international purchasing and supply chain executive working for Honeywell Aerospace as Senior Director of Integrated Supply Chain. Mr. Wincel founded and served as president of LSC Consulting Group for 11 years—an independent professional consulting service specializing in lean supply chain management. Mr. Wincel was also an adjunct professor of supply chain management for Grand Valley State University's MBA and undergraduate programs. In 2004, Mr. Wincel released his first book through Productivity Press, titled *Lean Supply Chain Management: A Handbook for Strategic Procurement.* Mr. Wincel has a BS and MBA from Michigan State University, and an MA from Loyola University–Chicago, and a doctorate from Graduate Theological Foundation. He has been selected as a "Pro to Know" by *Supply & Demand Chain Executive Magazine* in 2002, 2004, 2007 & 2012, and is a frequent contributor to many professional and industry publications. Mr. Wincel has also been a featured presenter at the University of Michigan Conference on Lean Manufacturing, the National Manufacturing Week Conference, the Institute for Supply Management, The Institute for International Research, and has been a repeat panelist on the ISM national satellite education series.

Thomas J. Kull is an assistant professor of supply chain management in the W.P. Carey School of Business at Arizona State University. Along with 13 years of industry experience in the steel processing, automotive, and construction equipment industries, he also holds a PhD in operations and sourcing management from Michigan State University. He researches behavioral issues in operations and supply chain management, as well as supply chain risk issues. He has published in the *Journal of Operations Management*, *Journal of Supply Chain Management*, *Decision Sciences Journal*, *IEEE Transactions on Engineering Management*, *European Journal of Operational Research*, and *International Journal of Operations and Production Management*. He serves on numerous editorial review boards, including *Journal of Operations Management*, *Decision Sciences Journal*, *Journal of Business Logistics*, and *Journal of Supply Chain Management*. He is a member of the Decision Sciences Institute, Academy of Management, and Production and Operations Management Society.

1

Introducing the Lean Culture

INTRODUCTION

Lean manufacturing (LM) and Lean systems surged onto the shores of the United States almost like a ninja from the East. The American automobile industry and other U.S. manufacturing sectors were having their lunches eaten by the Japanese, the same Japanese that just a few years earlier had goods that were synonymous with poorly made, cheap, throwaway products. Suddenly, almost out of nowhere, quality, reliability, and design were associated with these companies.

What was the secret that allowed the Japanese to become "overnight" successes? The answer would slowly emerge when books like K. Womack, Jones, and Roos' *The Machine that Changed the World* began to hit the bookstores and executives' desks in the early 1990s.[*] The successes of companies such as Toyota began to make some sense, albeit they were not fully understood or appreciated. For the next two decades, the American companies would chase this ethereal idea of LM. System after system, process after process, and tool after tool would be implemented in hopes of someday achieving this elusive success.

When many of the best efforts continued to fail and the Lean flavor of the month left a bitter taste in the mouths of employees and managers alike, the truth of Lean's success became painfully obvious: It is people who make Lean work, not tricks, processes, or elaborate automated equipment. But, now what? What does this mean concerning the cultures of the

[*] James P. Womack, Daniel T. Jones, and Daniel Roos, *The Machine that Changed the World: The Story of Lean Production*, Harper Perennial (November 1991).

people at the heart of the success of Lean? Clarity would be provided in Dr. Jeffrey Liker's book, *Toyota Culture: The Heart and Soul of the Toyota Way.*[*] A companion to Liker's 2003 book, *The Toyota Way*, Liker and Hoseus began to move Lean practitioners from thinking about process and tools only to building those process and tools around the culture from which the workers would utilize them. What Liker (and others) had finally discovered was that despite the efforts of U.S. manufacturers, even their best efforts, sustainable progress was not being made. In fact, Liker began writing about the efforts of these companies eleven years earlier yet perhaps had yet to see any successes even remotely comparable to those of the perennial powerhouse Toyota.[†]

COUNTERCULTURAL INFLUENCES

In the West, it is particularly difficult to wrap our business minds around the "countercultural" effectiveness and sustainable implementability of LM techniques with the cultural "persona" that usually indicates success. Our expected perception of the successful businessperson would include someone highly assertive, assured, risk taking, and the like. Yet, as we will explore in more detail, research data showed that LM is most effective in countries where motivational orientation is based on high avoidance of uncertainty (low risk taking), low assertiveness, low future orientation (works for and in the present moment), and low performance orientation.[‡]

In contrast, the difficulties in implementing and sustaining LM results in Western facilities are reflective of a strikingly different set of cultural norms. A striking example of this is reported in a 2010 *Economist* article detailing the woes faced by Toyota in its global expansion.[§] Often considered

[*] Jeffrey K. Liker and M. Hoseus, *Toyota Culture: The Heart and Soul of the Toyota Way*, McGraw-Hill, New York (2008).

[†] Dr. Liker has been writing about successful Lean practitioners since his 1997 book, *Becoming Lean: Inside Stories of U.S. Manufacturers*, Productivity Press, New York (1997).

[‡] Kull, Thomas J., Tingting Yan, and John G. Wacker, RP137—Cultural Influences on Lean Manufacturing Effectiveness (Working Paper Revision 4, January 2, 2011, p. 2).This was based on survey data from 24 countries and 1,400 manufacturing facilities. These surveys indicated that effective results in Lean manufacturing are sensitive to and subject to individual country-level cultural norms.

[§] The Machine that Ran Too Hot; Toyota's Overstreached Supply Chain *Economist*, 394 (8671), 74 (2010).

by many as the undisputed champion of Lean systems, Toyota's global expansion outpaced its (and its suppliers') ability to culturally adapt to a greater diversity of cultural variations. This "out-of-step" sequence with suppliers is more common and even more heightened moving through the supply chain tiers as resources and sophistication diminish. This perhaps is no new news, as the Toyota-GM experiment in California (New United Motor Manufacturing Inc., better known as NUMMI) and the GM-Suzuki JV in Canada (Canadian Automotive Manufacturing, Inc., better known as Cami), demonstrated early on that despite Toyota's expertise in LM and GM's willingness to acquire these skills, their effectiveness was limited to the cooperation and acceptance within the Western labor mindset. Cami particularly was subjected to (or even the victim of) labor's unwillingness or inability to integrate LM into the cultural norms of the Canadian automotive worker psyche. Despite the size of the effort or the amount of resources (financial or otherwise) given toward the implementation and integration of Lean, the workforce possesses the ability to derail the effort, to foil the ability to "embed" Lean culture and LM efforts into daily production activity.[*] While the ability (and often willingness) of labor to subvert the engraining of LM and Lean thinking into the production systems is often anticipated in the West, this is a lesser known (or even an unknown) phenomenon in the East. A workforce disconnected (from the cultural integration) of LM and its success possesses perhaps the greatest ability to disrupt the embedding of LM into the day-to-day activities of a manufacturing operation.[†]

Despite the proclivity to immediately "blame" the workforce for the absence of significant improvements in quality and productivity, the reality is a much more complicated picture. Employee responses to LM efforts are shaped culturally and geographically—both inter- and intranationally. There exists a great diversity in these national, social, and ethnic beliefs, practices, and values. As such, it is these differences that may point to the lack of uniformity in the success of LM in various regions and geographies around the world.[‡]

[*] J. Rinehart, C, Huxley, and D. Robertson, Worker Commitment and Labor-Management Relations under Lean Production at Cami, *Relations Industrielles-Industrial Relations* 49 (4), 750–775 (1994).

[†] A. Robinson and D. Schroeder, The role of front-line ideas in Lean performance improvement, *The Quality Management Journal* 16 (4), 27 (2009).

[‡] R.J. House, P.J. Hanges, M. Javidan, P.W. Dorfman, and V. Gupta, Eds., *Culture, Leadership, and Organizations: The GLOBE Study of 62 Societies*, Sage, London (2004).

The opposition of the typical Western perspective of LM to that from other parts of the world and in other cultures is that *in practice*[*] the Western perspective is that LM is a set or series of practices, things that are to be done, or the way in which things are done. Outside Western manufacturing, LM is equally or more importantly seen as a way of thinking, even perhaps a business philosophy. This is true despite well-respected practitioners and academicians in the West who write and teach to the contrary.[†] There continues to be an increased recognition of this "tension" between practice and intention. An analogy to this dichotomy could be the differences between ethics and morality, with ethics the ideology of the desired way things should be and morality the behavior in how the ideology is played out.[‡] In his 2009 book, *Toyota Kata: Managing People for Improvement*, University of Michigan professor Mike Rother described this idea by writing about a generally accepted core belief in LM that the implementation effectiveness is not limited by the desire or capability of the workforce but by the processes. Essentially, he related the belief that workers want to perform well, and effectively implement LM, but are limited by how well the system is designed.[§] In the West, there seems to be a difference between this general desire and the cultural realities that shape convictions and thought patterns. The root of the inability to connect the general desire for good outcomes and the expressed reality may be connected to the Western management teams charged with implementation. While it is generally held that the LM systems themselves (the practical application of tools/techniques) limit the effectiveness of a system, the workforce tends to take the blame (as previously). This "blame game" never allows the ideological shift to take hold; as a result, companies (or facilities/operations) that continue to blame the employees for system failure will continue to have failed expectations of LM mindset/ideology.[¶]

---

[*] I emphasize "in practice" because most LM education and implementation efforts stress the fundamental "cultural" and ideological changes necessary to achieve significant and sustainable improvement.

[†] Jeffrey K. Liker, *The Toyota Way*, McGraw Hill, New York (2003).

[‡] Jeffrey P. Wincel, *Defying the Trend: Business Ethics and Corporate Morality from a Faith Perspective*, God's Embrace Press, Holland, MI (2010), p. 12.

[§] Michael Rother, *Toyota Kata: Managing People for Improvement, Adaptiveness, and Superior Results*, McGraw-Hill, New York (2009).

[¶] J. Liker and M. Hoseus, Human resource development in Toyota culture, *International Journal of Human Resources Development and Management* 10 (1), 34 (2010).

THE DNA OF THE LEAN MANUFACTURING SYSTEM

For the purpose of understanding the cultural effect on the successful implementation of LM in various global and national cultures, Dr. Kull's research used the existing LM literature to develop what he called "LM values" and utilized eight different dimensions to categorize these values. These eight dimensions are unfolded over the next few chapters to describe the starting basis from which the cultural identifiers will be applied. Overlaying these "values" to the cultural dimensions of GLOBE (Global Leadership and Organizational Behavior Effectiveness), a picture will arise that "hypothesizes" or predicts the impact of country influence on LM implementation effectiveness. Finally, data from the Global Manufacturing Research Group (GMRG) are utilized to further validate the cultural impact on LM effectiveness, particularly across three specific cultural dimensions: (1) high uncertainty avoidance; (2) low assertiveness; and (3) low future orientation. As more fully explored in further chapters, the third dimension on low future orientation was a surprising cultural influence; data showed a greater effectiveness (and sustainability) of LM in organizations where there were fewer forward-looking expectations. This aspect is somewhat surprising and even perhaps a bit confusing as most practitioners view the business approach particularly in the East as extremely forward looking.

Unraveling the interplay between culture, tools, and LM effectiveness essentially reveals what Harvard Business School professors Steve Spear and Kent Bowen described as the "DNA" of the LM system.* These perspectives continue to be developed in current literature by a number of LM thought leaders, teachers, and academicians.†

The research of Kull et al. and, as an extension, the purpose of this management translation proposes to (1) clearly define the "global" LM values in the eight specific categories described previously; (2) show the ways in which the cultural-specific values influence and effectively determine the effectiveness of LM implementation and sustainability; and (3) provide

* Steven Spear and H. Kent Bowen, Decoding the DNA of the Toyota production system, *Harvard Business Review* 77 (9&10), 97–106 (1999).

† Some of these works include the following: Steven Spear, The Toyota Production System: An Example of Managing Complex Social/Technical Systems. 5 Rules for Designing, Operating, and Improving Activities, Activity-Connections, and Flow-Paths, unpublished DBA, Harvard University; Liker & Hoseus, *Toyota Culture*; Rother, *Toyota Kata*; and, John Shook, How to change a culture: Lessons from Nummi, *MIT Sloan Management Review* 51 (9), 1527–1546 (2010).

actionable information to managers charged with the task of introducing and implementing a Lean environment within their operations.

COMPLEX AND ORGANIC

Much of the professional and academic material seems to treat the Lean system disciplines as a "mechanistic" set of approaches that can be implemented and essentially force change to occur within a closed system. This is seen in various books and journals that describe finite steps or tools with a singularity of results common in essentially every implementation or application. There tends to be an assumption of a linear nature to the practice of Lean as well: Steps occur sequentially and logically from the preceding action, almost formulaically. What this "descriptive model" perspective fails to recognize is that despite our best efforts, the world is a complex place where individual results arising from the same approach do not necessarily yield identical outcomes. Also, the changes within this complexity should assume an "organic" response or growth; that is, the outcome cannot be forced but must emerge from the system itself.

For example, an underlying principle of small incremental change within a Lean system, which is generally an accepted part of Lean practice, is based on the premise that the outcome cannot be fully predicted. Changes are "small" not only so they are more manageable, but also so they can be reversed if the outcome is not as anticipated or as desired. If the world were mechanistic, as some authors would have you believe, there would be no uncertainty of the outcome only ineffective (or defective) implementation. In an organic complex world, the small change approach accommodates a greater level of uncertainty in the outcome reflective of the organic nature of the action—one that develops naturally or gradually without being forced or contrived.

This view of a complex and organic world in which manufacturing (and manufacturing systems) takes place can either be a facilitator or an inhibitor to the implementation of Lean practice. Nearly every author or practitioner will tell you that Lean implementation begins with a change of corporate culture, a change from a "traditional" mindset or view to one oriented toward Lean. They usually make this sound as if it as simple as changing your mind and simply doing it. Yet, this view also presumes a mechanistic view in which there is one action and one reaction: You change your mindset and are ready for Lean. Perhaps nothing could be further from the truth. The fundamental shift in mindset to a Lean culture is perhaps the most difficult step in the process. In many ways, it seems to be implied that the culture must come first before the components of the Lean system can be effective. We propose that this is far from the true reality of Lean approaches.

Western business planners bring with them a misunderstanding of the approaches in the East regarding how broader business fundamentals are integrated with Lean. In the late 1970s and early 1980s as U.S.-based manufacturing began to radically decline from the quality and efficiency gains of foreign competitors, executives began to believe that the shortness of the U.S. planning horizon was to blame. The focus on quarterly and annual performance seemed to be counter to the longer-term horizon (multiyear) of the new foreign competitors. This was an oversimplification and a critical misunderstanding of what the new competitors were really doing. While there indeed did exist a multiyear (or even multidecade) business plan of foreign competitors, it was not one made up of static discrete elements. It tended to be one with a long-term objective (or "north star") dominated by a view in which supporting practices would be implemented to deliver the results. This is the perspective of Lean within the broader business. There is not necessarily a "grand plan" that delineates every action and desired outcome, but a guiding principal to which the Lean efforts will be applied and evaluated.

So, returning to the cultural transformation of an organization seeking to implement Lean, by applying the complex and organic perspective with the guiding principal view, Lean culture can "grow" as Lean techniques are implemented and mastered. The use of Lean tools themselves creates the backdrop from which experience (success or failure) reinforces the culture of Lean, permitting the cultural shift to develop organically within the organization. Perhaps this is best understood in the adage that form follows function. Typically associated with modern architecture, the principle implied is that a building (or any object) should be primarily based on the intended function or use of the building or object.* In a like way,

* Louis Sullivan, *The Tall Office Building Artistically Considered*, Lippincott's Magazine, 1896. Louis Sullivan was the developer of the modern skyscraper design in Chicago. He was also mentor to Frank Lloyd Wright. The actual phrase used by Sullivan was "form ever follows function." The original magazine scan is available here: http://archive.org/details/tallofficebuildi00sull. Viewing is available online at these websites: http://www.scribd.com/ and http://academics.triton.edu/faculty/fheitzman/tallofficebuilding.html

the *form* Lean culture should follow is its *function*. Perhaps if the mechanistic and linear view of inverse, "function following form," in the world of Lean systems were to be slightly modified, the success of both would increase. As we move throughout the chapters in this book, we examine the research data and practical application of the form of Lean systems following its function particularly as it relates to global cultures.

QUESTIONS TO CONSIDER

1. Are cultural determinates of LM implementation clearly differentiated from tool deployment?
2. Is the success or failure of implementation understood in the context of system and culture?
3. Is the management team charged with deploying the LM system ideologically aligned with the system, or are these individuals governed by preexisting cultural bias?
4. Can a "traditional" company perspective (both management and labor view) grow to include some of the cultural drivers of LM effectiveness?
5. How can lessons learned from other cultural contexts be applied to the Western perspective?

2

Developing Lean Values

LEAN MANUFACTURING COMPONENTS

To understand and appreciate the cultural aspects of Lean manufacturing (LM), it is helpful to be reacquainted with the essential aspects or components of LM.* These six components are flow (FLOW), total preventive maintenance (TPM), employee involvement (EMP), pull systems (PULL), reduced setup times (SETUP), and the use of statistical process control (SPC). Each of these components affects and is affected by the cultural aspects and LM values in the manufacturing environments. The components are interrelated as well.

FLOW and PULL generate the continuous movement from raw material to customer; FLOW establishes the mechanisms, and PULL matches production with demand. Quick setup times aid the FLOW and PULL interaction by maximizing productive uptime. Similarly, SPC and TPM ensure stabilized and constant processes, the processes that are monitored by and acted on through employee intervention and involvement (EMP).

As we examine the cultural effects on LM, it is necessary to consider them along these dimensions of LM components. Table 2.1 shows the interrelationship between the cultural values (left side), the LM components (top), and the LM value (right side).

* R. Shah and P.T. Ward, Defining and developing measures of Lean production, *Journal of Operations Management* 25 (4), 785–805 (2007).

TABLE 2.1

Relationship between Value Dimensions and Lean Manufacturing

Cultural Values	Description	FLOW	Total Preventive Maintenance	Employee Involvement	PULL	Quick Setups	Statistical Process Control	LM Values
1. Truth: the basis of truth and rationality in the organization	Whether decisions are based on facts and data or intuition and experience	Flow rate is a calculated takt time that helps indicate waste in time and space to drive factory layout	Collect data and facts while operating equipment; adjust maintenance procedures if there is deviance between actual and expected	Experience and intuition of shop floor employees important; scientific method for improvement with teacher guidance	Production determined by visual signal of demand from the next station	Minimize changeover time and cost; adjust setup procedures if there are unexpected outcomes	Minimize process variation by analyzing data; control charts uncover special and common causes of variation	Base decisions on visible, firsthand facts and problems in a world that is difficult to predict
2. Time: the nature of time and time horizon	The time horizon for planning and goal setting and the urgency and timeliness in daily operations	Adhere to takt time in specified flow paths; long-term ideals guide daily improvements in small steps	Long-term vision guides short-term "target conditions"; follow specified procedures; frequent do-check cycles	Improve the activity while performing it; frequent and fine-grained diagnosis and problem solving	Production responds quickly as initiated by current demand instead of by planning	Long-term vision of zero setup time guides daily improvement; follow specified setup procedures, frequent problem solving	Long-term commitment for removing causes of process variation; immediate actions to defects	Urgent timeliness toward production and quick improvements are guided by long-term ideals

3. Motivation: the desires and motivations of people	Whether processes or people are the cause for poor performance	Improving factory layout and providing the right resources will increase productivity	Operators willing to do maintenance job; improve the way of doing maintenance	Operators willing to help improve the production system if involved in decision making	Workers monitor and adapt their behaviors according to signals; inventory reduction good to reveal problems	Poor processes cause long setup time; shorter setup time boosts work morale	Remove special and common causes of process variation	Problems are systemic and help reveal issues that people intrinsically are challenged to solve
4. Variance: stability versus change/ innovation/ personal growth	The aversion or desire for change/ improvement, incremental or radical change	Cells are flexible and evolving, with workers moving wherever needed and embracing continuous, stepwise improvement	Expand technical skills of personnel to do maintenance; use standards as reference for improvements	Standardized procedures for employee involvement; employees learn new skills and assume more roles	Incrementally reduce buffer inventory to realize an adaptable pull system	Breaking old habits critical for reducing setup time; use standards as reference for improvements	Keep processes within control; process capability improvement is never ending	Continuous, stepwise improvements prevent entropy, ensure stability, and focus on learning by doing

continued

TABLE 2.1 (continued)

Relationship between Value Dimensions and Lean Manufacturing

Cultural Values	Description	FLOW	Total Preventive Maintenance	Employee Involvement	PULL	Quick Setups	Statistical Process Control	LM Values
5. Vocation: orientation to work, task, and coworkers	The degree that productivity is important versus social relationships at work	Enable continuous flows through minimizing work-in-progress inventory and delays	Minimize equipment downtime to improve productivity, maintenance-production collaboration important	Employee involvement and collaboration improves decision-making quality	Minimize work-in-progress inventory to improve responsiveness	Reduce setup times to increase system productivity	Supply defect-free units through improving process capability	Productivity needs continuous monitoring with distinct work duties among operators and leaders
6. Cooperation: isolation versus collaboration	Whether working together improves performance	Collaboration is necessary for continuous flow	Voluntary small-group activities aid cross-functional collaboration	Manager-operator collaboration helps consensus decision making	Production initiated by kanban	Collaboration between setup experts and operators important	Teams support improvement activities; individual actions triggered by unexpected chart signals	Entire workforce is interdependent and must cooperate using guidelines to solve problems

7. Control: power, coordination, and responsibility	The “tightness” of control and usage of formal rules and standardized procedures	Standardized connections and detailed routines	Standardized maintenance procedures	Employees routinely involved in improvement according to roles, formal procedures, and programs	Production is strictly controlled by kanban cards	Standardized setup procedures	Operators take actions whenever processes are out of control; standardized monitoring and analysis procedures	Detailed routines create standards that temper irrationality and uncertainty while channeling human capabilities
8. Focus: organizational orientation and focus	If innovation is defined from within the organization or from outside	Self-benchmark for zero internal waste in time, space, and information	Minimizing internal equipment downtime	Internal employees are source for system improvement and problem solving	Production is always initiated to feed customers’ current demand	Short setup times enable fast response to customers’ requirements	Improve capability of internal processes	Self-benchmark to improve service performance using internal leader guidance

ORGANIZATIONAL VALUES

Frequently, the idea of any sort of values as an operating imperative in business operations seems foreign to Western ears. Values, or values and morals, are typically viewed as essentially divorced from the business environment.* Yet, there is a clear correlation between the success of organizational management (OM) efforts and the organization value underpinning those efforts. Kull's research pointed to a number of OM-based research articles that showed the clear correlation between the complimentary organizational values and the successful implementation of the OM initiatives.†

Organizational values, or values of any sort, are based in culture and as such may vary throughout the world. Sometimes described as culturally based deontology, the values (derived from culturally based ethics and morality) reflect the particular culture from which they rise.‡ While there generally tend to be cross-cultural values, those related to work can vary greatly with culture. These variations determine the importance and appropriateness of various organizational values.§ The resulting implementation of OM practices is skewed in various ways by the determined level of appropriateness. The various implementations may either support or resist the underlying efforts depending on the level at which they are deemed to be congruent with cultural norms and values. Where the underlying cultural values are incompatible or unsupportive of the initiative, an organization may continue to implement the initiative for the sake

* Wincel, *Defying the Trend.*

† This research includes J.R. Detert, R.G. Schroeder, and J.J. Mauriel, A framework for linking culture and improvement initiatives in organizations, *The Academy of Management Review* 25 (4), 850 (2000); B.B. Flynn, R.G. Schroeder, and S. Sakakibara, A framework for quality management research and an associated measurement instrument, *Journal of Operations Management* 11 (4), 339–366 (1994); K.E. McKone and E. Weiss, TPM: Planned and autonomous maintenance: Bridging the gap between practice and research, *Production and Operations Management* 7 (4), 335–351 (1998); Spear and Bowen, Decoding the DNA; R.E. White, J.N. Pearson, and J.R. Wilson, JIT manufacturing: A survey of implementations in small and large U.S. manufacturers, *Management Science* 45 (1), 7–20 (1999).

‡ For more information on deontological ethics, see *Stanford University Encyclopedia of Philosophy*, http://plato.stanford.edu/entries/ethics-deontological/ accessed December 14, 2011.

§ C. O'Reilly and J. Chatman, Culture as social control: Corporations, cults, and commitment, in B.M. Staw and L.L. Cummings, Eds., *Research in Organizational Behavior*, pp. 157–200, JAI Press, Greenwich, CT (1996).

of appearance.[*] The implementation of these efforts is done superficially and is done so conditionally—that is, redundantly—as an overlay on the old system, which is not replaced.[†] Moreover, these new systems or initiatives are not seen as a complimentary add-on to the existing systems but often as damaging and hastily deployed.[‡] Often, where cultural values do not support the OM initiatives, employees see implantation as potentially subversive or born from company politics.[§] Highly unionized Western companies are replete with anecdotal observations of both the reality of and the suspicion of political motivations for the introduction of new organizational initiatives. For LM initiatives and implementation to be successful, the managing organizations have to be aware of *and* responsive to the underlying cultural values that *will* have an impact on success.

THE EIGHT LEAN VALUES

In their *Academy of Management Review* article (2000), Detert, Schroeder, and Mauriel provided eight dimensions of cultural influence on the values and beliefs that underlie management systems. From this work, Kull and Wacker (2010) expanded these values specifically to LM initiatives. Developing (really articulating) what could be described as LM values, Table 2.1 shows the way in which these factors and general values translate to LM initiatives.

Perhaps unsurprisingly, value 1 is truth. This value serves to describe facts and rationality that are present within an organization and the effect their

[*] E. Abrahamson, Managerial fads and fashions—The diffusion and rejection of innovations, *Academy of Management Review* 16 (3), 586–612 (1991); K. Aoki, Transferring Japanese kaizen activities to overseas plants in China, *International Journal of Operations and Production Management* 28 (6), 518–539 (2008).

[†] X. Koufteros and M. Vonderembse, The impact of organizational structure on the level of JIT attainment: Towards theory development, *International Journal of Production Research* 36 (10), 2863–2878 (1998).

[‡] J. Fucini and S. Fucini, *Working for the Japanese*, Free Press, New York (1990); S. Green, The missing arguments of Lean construction, *Construction Management and Economics* 17 (2), 133–137 (1999).

[§] D. Beale, *Driven by Nissan? A Critical Guide to New Management Techniques*, Lawrence and Wishart, London (1994); P. Turnbell, The limits to Japanization-just-in-time, labour relations and the UK automotive industry, *New Tech, Work and Employment* 3 (1), 7–20 (1988); and White, Pearson, and Wilson, *JIT Manufacturing*.

presence (or absence) has on the implementation effectiveness of LM (and other OM initiatives). Truth in this context relates to the decision criteria for action (or inaction) based on observable and firsthand facts. This value seeks to eliminate decision-making uncertainty typical in manufacturing environments (culturally driven, customer driven, demand driven, etc.).

Cultural Values	→	LM Values
1. Truth: the basis of truth and rationality in the organization		Base decisions on visible, firsthand facts and problems in a world that is difficult to predict

Lean value 2 is time. In this usage, time is the dimension in which opportunity exists as a resource as both the nature and horizon available. This would include the amount of time available for decision making as well as the timeliness of the decision making. In LM implementation, this would mean the "urgent timeliness" toward production as well as the "quick improvements" aligned with the long-term and overarching ideals of a company.

Cultural Values	→	LM Values
2. Time: the nature of time and time horizon		Urgent timeliness toward production and quick improvements are guided by long-term ideals

Motivation is Lean value 3. Motivation does not necessarily reduce itself to compensation concerns but more toward intrinsic motivational characteristics. As an LM value, motivation includes the systematic identification of problems and implementation/effectiveness issues in such a way that the desire and motivations of employees are challenged to find resolution to these issues.

Cultural Values	→	LM Values
3. Motivation: the desires and motivations of people		Problems are systemic and help reveal issues that people intrinsically are challenged to solve

An easily recognizable LM element, variance, is Lean value 4. Here, there are two equally important dimensions of variance. The first is that which all Lean practitioners understand, the continual improvement within the Lean system to prevent entropy (uncertainty associated with random variables); to ensure process (and product) stability; and to focus on activity-based learning. The second dimension of variance is the attitude and cultural response of people to the continuum of change versus stability.

Cultural Values	→	LM Values
4. Variance: stability versus change/innovation/ personal growth		Continuous, stepwise improvements prevent entropy, ensure stability, and focus on learning by doing

Although not usually equated with Lean systems or LM, Lean value 5 is vocation. In LM culture, vocation does not emerge from a drive, desire, or ambition from within the worker but is defined from the outside in. Vocation in this sense is determined by the distinctly identified work duties necessitated by the need for continuous performance monitoring of productivity levels and improvements. Essentially, vocation is the cultural (and operational) orientation toward the work being completed, toward the individual tasks within the work, and toward coworkers engaged in the common tasks and work. The vocational aspects and view of LM apply equally to the workers as well as the Lean system leaders.

Cultural Values	→	LM Values
5. Vocation: orientation to work, task, and coworkers		Productivity needs continuous monitoring with distinct work duties among operators and leaders

Cooperation is Lean value 6. This value takes on particular importance and relevance culturally, especially in those areas in which independent work structures tend to be more culturally accepted and expected. There are significant variances in how this might appear in different cultural contexts. In certain situations, isolation would describe the work preference due to less emphasis and expectation of interaction. While appearing similar, independent work (as opposed to isolated work) would exist where individualism and individual results are rewarded. Within LM implementation, neither form of individual work can predominate as LM requires interdependence and active cooperation among the individual members of the workforce.

Cultural Values	→	LM Values
6. Cooperation: isolation versus collaboration		Entire workforce is interdependent and must cooperate using guidelines to solve problems

Lean value 7 is control, and again can be a culturally charged or determined value. Control includes the various aspects and meanings of power for which cultural perspectives may have great importance. Some views of power may indicate the ability to manage systems and the people operating those systems, whereas other cultures may view power as primarily a characteristic for self-determination and personal success outcomes. In our usage as an LM value, control implies the exercise of authority over the systems and practices of LM to influence the desired outcomes. This view of control includes the development, implementation, and use of detailed routines and standards that are designed to moderate or minimize the irrationality or uncertainty of people and their interface with process while effectively channeling human capabilities.

Cultural Values	➜	LM Values
7. Control: power, coordination, and responsibility		Detailed routines create standards that temper irrationality and uncertainty while channeling human capabilities

The last of the LM values, the eighth, is focus. Culturally, focus relates to the way in which LM innovation is determined—whether from inside or outside the organization. Focus is a "self-benchmark" for which internal leader guidance drives to improve service performance. LM is generally understood to have been developed or employed through an internal structure that included a mentor-mentee relationship. It is out of this relationship that LM expertise is developed internally to ensure Lean operations.

Cultural Values	➜	LM Values
8. Focus: organizational orientation and focus		Self-benchmark to improve service performance using internal leader guidance

LEAN VALUE CONSISTENCY

Highly Consistent Values

In understanding the translation from cultural values to LM, Table 2.1 indicates the degree of significance between the cultural values and LM components (Table 2.1, highlighted in gray). Those values with greater relevance (shown with more highlighted areas) suggest a greater influence of culture on the Lean component. There are four values that were considered by the researchers as "highly consistent": time (LM2), motivation (LM3), variance (LM4), and control (LM7). The remaining four values are considered "partially consistent": truth (LM1), cooperation (LM5), vocation (LM6), and focus (LM8).

LM2 Time

Beginning with the highly consistent characteristics, Table 2.1 shows LM2 Time has high relevance with FLOW, TPM, EPM, SETUP, and SPC. Across these components, cultural value has a marked impact across a broad range, reflecting a consistent theme of "urgent timeliness." This theme is a common cultural view in which long-term ideals guide production and production improvement, especially quick improvements. This quick problem solving is driven by the pace of adherence to customer demand ("takt" time) and at all levels of the organization.* This is in keeping with the "true north" of LM, that is, achieving one-piece, sequence-to-demand FLOW, with zero defects or waste and providing for employee security. This value is embodied in all LM components. For example, FLOW requires adherence of all processes to scheduled/planned production rates. Similarly, the frequent checks of TPM must be made by trained employees; PULL stresses short runs capable of accommodating frequent, unpredictable, or irregular changes of customer demand schedules. Finally, SETUP works toward achieving a waste-free, zero-setup-time production process. There is universality in the various cultural perspectives regarding time; a sense of urgency and quest for ideals embody this LM value.

LM3 Motivation

All six LM components commonly assume that the underlying problems, which LM techniques seek to address and improve, are systematic, and that they expose fundamental issues that people are intrinsically motivated to address and improve. It is this second consistent LM value, Motivation (LM3), that spans the continuum of cultural perspectives and LM components. LM (and more broadly Lean production) views the manufacturing process as a series of "working hypotheses"; operators continually test and

* Rother, *Toyota Kata* (2009).

move toward the "best production methods."[*] Intrinsic motivation compels employees to identify existing manufacturing issues/problems. It is these problems, and the solutions stemming from them, where improved manufacturing methods are found.[†] Employee participation and involvement (EMP) as a LM value is the facilitating mechanism that channels this motivation and desire to do a good job, capturing the employee knowledge regarding what causes poor process performance and why.

Traditional Western manufacturing philosophies tend to diverge from intrinsic motivational thinking, especially with regard to process improvement motivation. As a fundamental expectation, LM assumes continuous FLOW as an operational hypothesis where properly designed manufacturing processes exist with fully trained and resourced employees. It is these processes, resources, and employees that enable continuous FLOW to exist.[‡] Typically, Western perspectives on FLOW and process and equipment maintenance (particularly in a unionized environment) rely on a reward-and-punishment approach. In contrast, LM does not see this approach as an appropriate (or effective) motivator. Rather, it sees intrinsic motivation with the appropriate support and resources as the cultural stimulus that causes workers to want to maintain manufacturing equipment while producing.[§] As a LM value, motivation assumes that human capital (people) is valued as solutions to systematic problems.

LM4 Variance

The third consistent LM value across cultures is variance. Despite the tendency to view variance as change, LM views variance as stability, even while it is made increasingly possible through change. Change in the LM environment is valued on the condition that it is done in a purposeful, planned, and stepped fashion. Change as seen in a continuous and conscious fashion provides measured improvement with process stability with process entropy (uncertainty) prevented. This limited perspective of change in LM is based on the understanding of the existence of high

---

[*] Spear and Bowen, Decoding the DNA (1999).

[†] Shook, How to change a culture (2010).

[‡] S.D. Treville and J. Antonakis, Could Lean production job design be intrinsically motivating? Contextual, configurational, and levels-of-analysis issues, *Journal of Operations Management* 24 (2), 99–123 (2006).

[§] K.E. McKone and E. Weiss, TPM: Planned and autonomous maintenance: Bridging the gap between practice and research, *Production and Operations Management* 7 (4), 335–351 (1998).

complexity within manufacturing processes and the need to fully understand the cause and effect prior to change implementation.*

Process uncertainty (process entropy), even in established LM environments, requires constant monitoring and process correction. Cellular manufacturing, frequently an aspect of LM workplaces, is particularly dependent on process monitoring and correction flexibility, as well as an "evolving" organization that embraces continuous improvement philosophy.† In this and other LM environments, continuous monitoring and adjustments facilitate PULL system ongoing optimization with continuous reductions in production inventory, time, and manufacturing capacity buffers.‡ Variance affects SETUP in the need to redesign work areas *and* the control of production, or even the entire production organization model, to effectively manage process entropy.§ Variance control in TPM and FLOW is observed in the cross-training and expanded technical skills of manufacturing personnel, enabling maintenance work and job flexibility and allowing resource deployment where and as needed. On the variance dimension, SPC is observed in the emphasis on the "never-ending" improvements in process capability.¶ Ultimately, on this "consistent" cultural dimension, variance describes the systematic approach to "breaking old habits" in pursuit of operational excellence in the LM environment.

LM7 Control

Finally, the fourth fully consistent LM dimension across the various cultural values is control, LM7. As LM operational philosophy assumes *subjective* human judgment in assessing process entropy (uncertainty), employee assessment and intervention are supported by standard routines that reinforce specific patterns of behavior. These routines and behavior patterns become the reference points from which improvement actions are taken.** As an extension of this, all eight LM components rely on the

* Rother, *Toyota Kata* (2009).
† N.L. Hyer, K.K. Brown, and S. Zimmerman, A socio-technical systems approach to cell design: Case study and analysis, *Journal of Operations Management* 17 (2), 179–203 (1999).
‡ K. Schultz, D. Juran, J. Boudreau, J. McClain, and L.J. Thomas, Modeling and worker motivation in JIT production systems, *Management Science* 44, 1595–1607 (1998).
§ M. Diaby, Optimal setup time reduction for a single product with dynamic demands, *European Journal of Operational Research* 85 (3), 532–540 (1995).
¶ Rother, *Toyota Kata* (2009).
** Ibid.

existence of formal rules and standardized (and formalized) procedures to provide coordination of and between the various production processes.*

The standardization of rules and work/process procedures is seen in PULL and FLOW in the use of kanban. Typically, production velocity and mix are "strictly controlled" by formalized (and specific) kanban demand signals.† Typically, or frequently misunderstood in Western application, kanban in this perspective is clearly more important than simply as a means of inventory flow and control. Control as an LM component drives SPC and TPM actions by the use of highly detailed procedures that guide specific responses and actions in the face of "out-of-control" situations. These detailed procedures are the basis for quality assurance and breakdown prevention actions. In these aspects, there can be a misunderstanding of the meaning and purpose in employee involvement (EMP). Again, where Western perspectives would suggest a high level of discretion and autonomy, LM anticipates the regular and routine involvement of employees with limited autonomy. This limitation, however, does not lessen the importance of the employee involvement, but rather emphasizes the differences in the cultural view of the nature of involvement. This has been described by saying that, "Individual autonomy plays hardly any role of importance, and even improvement activities and creativity have been standardized."‡ Perhaps the cultural differentiation is understood in that although employee autonomy is greatly reduced in a LM work environment, the purpose is not to lessen higher-function decision-making capabilities but to establish operational discipline where actual performance can be assessed against planned performance with improvement activities specifically focused to bridge any divergence.§ When one considers the potential effects of operational control and efficiency as well as product quality in an environment of uncertainty (entropy), the control mindset and mechanisms are valuable aids in containing and responding to the uncertainty.

A useful analogy in understanding the apparent disconnect of autonomy in decision making and action might be to consider the practice of martial

* Spear and Bowen, Decoding the DNA (1999).

† Y. Sugimori, K. Kusunoki, F. Cho, and S. Uchikawa, Toyota production system and kanban system materialization of just-in-time and respect-for-human system, *International Journal of Production Research* 15 (6), 553–564 (1977).

‡ W. Niepce and E. Molleman, Work design issues in Lean production from a sociotechnical systems perspective: Neo-Taylorism or the next step in sociotechnical design? *Human Relations* 51 (3), 259–287 (1998).

§ Rother, *Toyota Kata* (2009).

arts. From an outside observer, it may appear that an accomplished martial artist acts completely autonomously and creatively in responding to outside threats. This, however, is not typically the reality of the situation. The progression from an inexperienced 10 kyu (級) novice (unbelted or white belt) to a seasoned first dan (段) black belt includes the practice and repetition of stylized forms called *katas* (型). Katas are a series of forms or choreographed moves/patterns designed to provide a practiced response (or defense) in a particular situation. The purpose of these katas/forms is to engrain the response in memory, particularly in muscle memory, so that they become "automatic" when needed. "Testing" for rank promotion is both a function of time of practice and demonstration of knowledge *and* mastery of the forms. By the time a practitioner achieves his or her black belt rank, the person will have rehearsed the katas literally thousands of times and employed them during sparring practice. The strength of this analogy also includes the response time differential from a novice white belt to a more expert black belt. Studies have shown that the response and reaction times of practiced martial arts students can be ten times that of the average response time. In LM, the response time to operational uncertainty is improved through experience and the practice of "automatic" responses. Finally, most forms of martial arts use even simpler forms called *wazas* (or *Kihonwaza*) as part of their testing and promotion requirements. Wazas are the very "basic techniques," often considered the foundation of the practice. Typically, only three to five basic moves, a waza is the building block of the entire practice, just as the individual elements of standardized work (or other LM basics) might be considered the building blocks of a flexible and robust process in LM.

Partially Consistent LM Values

Whereas the four culturally consistent values are fully applicable to all LM components, partially consistent LM values are not completely applicable to all LM components. These remaining four "partially consistent" values are truth (LM1), cooperation (LM5), vocation (LM6), and focus (LM8).

LM1 Truth

As the first of these dimensions, truth (LM1) may seem somewhat out of place as a business consideration, although it undoubtedly plays an important role as a cultural value. As an LM value, truth is understood as the basis for rationality within an organization. Decisions in the LM environment are based on visible, firsthand facts and problems. LM assumes that such facts and problems occur in a "difficult-to-predict" world. LM also assumes that this unpredictable world is highly interactive and constantly changing.[*] In this setting, rationality in the production environment requires firsthand factual information and the utilization of scientific methods over personal intuition as the basis for decision making in LM.[†]

With continuous FLOW as a basic LM component, truth as a value described previously permits achieving this objective through the required use of technical product specifications and operator skill sets as the basis of factory layout optimization over subjective preferences or opinions.[‡] In a similar way, SPC in process control and capability analysis depends on the factual use of process data in statistical analysis tools to determine or evaluate process control and stability. But, from a seemingly conflicting perspective, visual management of production coordination is valued over data analysis in the LM environment.[§] The contradiction seems to be that the truth of manufacturing coordination is based on apparently subjective evaluation of visual input. However, even visual management is based on data and informationally driven plant layout and organization. Ironically, the use of visual management makes corporate (or remote) monitoring of manufacturing or operational performance difficult without visiting the facility.[¶]

The LM EMP component intersects the truth value in somewhat of a counterintuitive way. As previously stated, LM relies on observable data and judiciousness in the decision-making processes. Employee involvement in the LM environment allows for, and even encourages, operator intuition in problem identification.[**] As Table 2.1 states, EMP-truth dimensions rely on employee experience and intuition, while scientific methods are applied to make adjustments and improvements. Likewise, FLOW initiatives require the *abstract* calculation of takt time, yet become the rational fact basis of waste elimination on the shop floor. These two

---

[*] Ibid.

[†] Spear and Bowen, Decoding the DNA (1999).

[‡] Hyer, Brown, and Zimmerman, A social-technical systems (1999); Shah and Ward, Defining and developing (2007).

[§] J.P. Womack and D.T. Jones, *Lean Thinking*, Simon and Schuster, London (1996).

[¶] Rother, *Toyota Kata* (2009).

[**] S. Sakakibara, B.B. Flynn, R.G. Schroeder, and W.T. Morris, The impact of just-in-time manufacturing and its infrastructure on manufacturing performance, *Management Science* 43 (9), 1246*M*/1257 (1997).

dimensions are only partially consistent with LM values. Overall, and despite these "exceptions," LM relies on facts and rationality as the basis to manage operational entropy.

LM5 Vocation

The next partially consistent LM value is vocation. An important feature in LM is the practice of continuous process monitoring. The essential purpose of this continuous monitoring is to ensure efficiency of maintaining established production rates. Successful, and repeated, adherence to discrete work tasks by production employees enables production rate adherence (and ultimately schedule attainment), with continuous process monitoring as the verification of this work. More than just an efficiency assurance measure, continuous process monitoring reflects the belief that production rates are a shared priority of operators, supervisors, and managers. Vocationally, the organization is *devoted* to maintaining production rates.[*]

As a result of this dedication and process monitoring of performance, increasing the level of productivity throughout the manufacturing process (throughout the entire system) results in greater priority than preserving personal social comforts. For example, FLOW and PULL initiatives and performance result in system efficiencies for which, through the reduction in inventory levels and time buffers, process variation and weakness are exposed. As a result, work environments may become or be considered more stressful as process variation is exposed.[†] Visual management techniques such as production boards and *andon* systems are specifically designed to openly reveal production problems to which workers are sensitive.[‡]

While visual systems provide for immediate feedback concerning performance, the employee sensitivity to the exposure of performance gaps can create an internal "tension" on the manufacturing floor. Likewise, LM discipline requires that operators follow specific work instruction and roles; these limitations may be perceived to create a demotivating work environment.[§] The adherence to specific work instructions and work roles

---

[*] Rother, *Toyota Kata* (2009).

[†] R. Conti, J. Angelis, C. Cooper, B. Faragher, and C. Gill, The effects of Lean production on worker job stress, *International Journal of Operations and Production Management* 26 (9), 1013 (2006).

[‡] Shook, How to change a culture (2010).

[§] Treville and Antonakis, Could Lean production (2006).

is not intended (necessarily) to limit individual worker initiative or motivation, but determined and enforced so that workers are aware and able to understand what is happening in the workplace.[*] This distinction in purpose and motivation is significant in that the general cultural perspective in LM is that self-motivation and pride overcome the potential negatives associated with structured work. For all these reasons, vocation as an LM value is viewed as highly valuing a "productive orientation" toward work roles.

LM6 Cooperation

The third partially consistent LM value is cooperation (LM6). On this dimension, LM adopts the perspective of full interdependence of the entire workforce. This interdependence requires collaborative participation among the employees utilizing LM-established guidelines for problem solving. In this way, the decision making is made via collaboration across departments and hierarchies. As a result, it is generally believed that decisions arising from this approach are superior to individual decisions (or stemming from individualism), as collaborative decisions integrate various organizational knowledge and knowledge bases to increase the quality of decision making.[†] Not only are decisions made cross organizationally and cross departmentally, but also LM purposely overlaps responsibilities between team leaders and team members.[‡]

This interdependence of teams, team members, and team leaders is clearly seen in the LM environment from beginning to end of the production process. For example, while production operators are dependent on "setup experts" for robust designs and implementation of rapid tool-change processes[§] (single-minute exchange of dies), the setup experts and designers are similarly dependent on the operators to collaborate with maintenance workers in deriving the solutions to equipment (and tooling) problems. In the heart of the manufacturing process, achieving

[*] Rother, *Toyota Kata* (2009).

[†] S. Ahmad, R.G. Schroeder, and K.K. Sinha, The role of infrastructure practices in the effectiveness of JIT practices: Implications for plant competitiveness, *Journal of Engineering and Technology Management* 20 (3), 161–191 (2003); B. Dankbaar, Lean production: denial, confirmation or extension of sociotechnical systems design? *Human Relations* 50 (5), 567–584 (1997).

[‡] Rother, *Toyota Kata* (2009).

[§] S. Shingo, *A Revolution in Manufacturing: The SMED System*, Productivity Press, Cambridge, MA (1985).

continuous product FLOW in a "healthy" work environment requires ongoing and extensive communication and dialogue between work cell operators and production supervisors.*

The development, placement, and use of standardized work with its highly detailed instructions and derivatives lessen the need for cooperation within and among workers and supervisors. PULL systems, for example, tend to reduce worker interactions by the use of kanban systems and techniques.† Perhaps contrary to popular understanding of the LM systems, and especially contrary to the manufacturing practices in the West, LM discourages informal cooperation in work cells.‡ Despite these seeming contradictions to the Lean value of cooperation, LM presumes an interdependency of the workers in the LM environment and values and encourages supportive, collaborative, and cooperative problem solving of its team members, supervisors, and managers.

Focus LM8

Wrapping up the partially consistent LM values is the LM value of focus (LM8). In LM, focus primarily refers to the self-benchmarking of process and performance to achieve improvements in customer service and satisfaction. Typically, internal benchmarking assessments are performed by the production operators with the guidance of internal leaders. However, in the LM environments across cultural settings, there is a mixed focus based on whether the success drivers are internal or external to the organization.§ This mix focus arises when one considers that the development and practice of LM emerged through the "mentor-mentee" relationship in which employees are trained as internal experts ensuring Lean operations, yet the measure of their success is the external customer satisfaction.

In their book *Lean Thinking,* Womack and Jones wrote that the continual pursuit of excellence in the LM operations is primarily driven

---

* Hyer, Brown, and Zimmerman, A social-technical systems (1999).

† J.P. MacDuffie and J.F. Krafcik, Integrating technology and human resources for high performance manufacturing: Evidence from the international auto industry, in T.A. Kochan and M. Useem, Eds., *Transforming Organizations*, Oxford University Press, New York (1992); R. McLachlin, Management initiatives and just-in-time manufacturing, *Journal of Operations Management* 15 (4), 271–292 (1997); Niepce and Molleman, Work design issues (1998).

‡ Rother, *Toyota Kata* (2009).

§ Detert, Schroeder, and Mauriel, A framework for linking culture (2000).

by an internal process of self-improvement.* Internal resources and decision-making initiatives are used to improve efficiency and productivity. This is seen in FLOW, TPM, EMP, and SPC, in which internal orientation and internal resources are used to improve total system performance by reducing all forms of manufacturing or operating waste; allowing for greater equipment effectiveness; creating an environment in which decision-making quality is better; and creating more capable production processes.† In contrast to the internal focus, PULL and SETUP are essentially externally oriented. These LM characteristics highlight that LM system success is based on (or is dependent on) being able to rapidly respond to external changes in demand irrespective or regardless of any internal limitations or constraints of overtime, previously established schedules, or employee stress.‡

LM itself is mixed focus. While primarily an internal process of self-improvement, LM also relies on external influences to achieve improvement. FLOW, TPM, EMP, and SPC are principally (or exclusively) internally focused. Internal resources are used to improve productivity throughout the manufacturing system through less waste, greater equipment effectiveness, better decision quality, and more capable processes. In contrast, PULL and SETUP rely on external information regarding changes in customer demand and the timely response to these changes to maintain LM success. Even the external components become part of and are governed by the internal means to achieve customer satisfaction. Fundamentally, LM values satisfy the external customer influence and demands through internal excellence by continual self-improvement.

---

* Womack and Jones, *Lean Thinking* (1996).

† P.J. Egbelu and H.P. Wang, Scheduling for just-in-time manufacturing, *Engineering Costs and Production Economics* 16 (2), 117–124 (1989); Hyer, Brown, and Zimmerman, A social-technical systems (1999); Koufteros and Vonderembse, The impact of organizational structure (1998); K.E. McKone, R.G. Schroeder, and K.O. Cua, The impact of total productive maintenance practices on manufacturing performance, *Journal of Operations Management* 19 (1), 39–58 (2001); and M. Rungtusanatham, J.C. Anderson, and K.J. Dooley, Conceptualizing organizational implementation and practice of statistical process control, *Journal of Quality Management* 2 (1), 113–137 (1997).

‡ A. Edström and Olhager, J., Production-Economic Aspects on Set-up Efficiency, *Engineering Costs and Production Economics* 12 (1-4), 99-106, (1987); and Gottfried, H. & Graham, L., Constructing Difference: The Making of Gendered Subcultures in a Japanese Automobile Assembly Plant. *Sociology* 27 (4), 611–628, (1993).

SUMMARY

Individual cultural values are translated into LM values by observing and understanding the translation from culture to LM through the application of the LM components. Examining the way in which the correlation between LM and culture bear on each component indicates which values are fully consistent and which are partially consistent. Within both the fully and the partially consistent values, there may also exist apparent contradictions of the applicability of the value. These contradictions tend to be a product of the cultural interpretation or experience of the way in which the characteristic is perceived. For example, the West may view the cultural practices of the East that limit individuality as contrary to employee satisfaction and participation. However, underpinning the Eastern view is the assumption of internal motivation and vocational desire that eliminates (or minimizes) individualism as a factor in employee performance and satisfaction. Understanding the way in which partially consistent values vary from culture to culture is an essential element to anticipate cultural resistance or misapplication of LM components.

QUESTIONS TO CONSIDER

1. Has your organization considered specific cultural values as part of its LM implementation?
2. How are the LM characteristics applied considering the cultural dimension?
3. Are there culturally based interpretations that affect the employees' and managers' perspective of the meaning of LM characteristics?
4. In what ways are team leaders and managers capable of addressing and empowered to address the cultural interpretations?
5. Where in your manufacturing organization is the success of LM performance struggling due to culture?

3

The Hypotheses of Lean Culture

DEVELOPING THE HYPOTHESES

Life in the Lean environment can often appear to be contradictory. For example, one of the hallmarks of Lean systems is the reliance on statistical data, standardized (i.e., documented) approaches, process control characteristics, and other *objective* standards. Yet, at the same time, Lean monitoring also relies on the visual factor, which could be considered *subjective* evaluations. This apparent contradiction exists only if one views Lean culture as an either/or proposition, that is, if objective and subjective measures are seen as mutually exclusive. In the Lean environment, both the objective and subjective, the statistical and the observed, are reflections of what is seen and perceived. Both shed light on the reality of the situation as it presently exists, and both aid in determining the appropriate actions (if any) that are necessary to reshape the present situation into a different future state.

Understanding the cultural effect on the implementation of Lean manufacturing (LM), especially the Lean supply chain aspects that intersect multiple or various cultural backdrops, is realized in the holding of simultaneous truths. To recognize and be able to act on the cultural differences, it is necessary to understand the way in which the cultural dimension and Lean dimension have been harmonized in the research and in this book. But, do not worry; there will not be an overabundance of academic-ese, only just enough to help managers understand the connections between the two and the methods of their connection. If you choose to skip past this chapter, be sure to circle back at the end as this approach will shed important light on how to effectively manage when exposed to and in new cultures.

HYPOTHESIS 1: LEAN PRACTICES ARE POSITIVELY RELATED TO PERFORMANCE

In the world of academia, hypotheses are derived as a result of observation, then tested to validate them. Lean thinking is also no stranger to hypothesis testing. In their *Harvard Business Review* article, Spear and Bowen (1999) described that in Toyota's Lean thinking, hypotheses are constantly being tested—that is, hypotheses about what is the "best" way to run a particular plant.* In this chapter, we similarly follow such Lean thinking by giving hypotheses about what is the best way to think of LM in different cultures. For our efforts, it may be simpler to state the hypotheses first and then summarize the basis for them.

Hypothesis 1 is *that Lean manufacturing practices are positively related to operating performance.* To describe this first hypothesis in everyday terms, LM practice increases operating performance in all cultures around the world. For those already embedded in this world, that is probably a statement similar to the one about motherhood and apple pie, one with no great mystery. Yet, this statement helps us reset the building blocks of our understanding of the elements of Lean systems. In fact, much of this hypothesis was laid out in the first two chapters; still, it bears repeating here.

Lean practices are effective and positively affect operating performance through "creating a streamlined, high quality system that produces finished products at the pace of customer demand with little or no waste."† As simple and straightforward as this may seem, Lean, unlike other production initiatives, creates this positive effect across a wide array of production and supply chain characteristics. Essentially, all of the elemental Lean characteristics can be directly correlated with specific outcomes and operational efficiency. For example, manufacturing cycle times are improved and inventory reduced by Lean *continuous flow* processes‡; production

* Spear and Bowen, Decoding the DNA (1999).

† R. Shah and P.T. Ward, Lean manufacturing: Context, practice bundles, and performance, *Journal of Operations Management* 21 (2), 129 (2003).

‡ Hyer, Brown, and Zimmerman, A socio-technical systems approach (1999).

efficiency is effectively maximized though greater equipment reliability with the implementation of *total preventive maintenance* (TPM). Greater equipment "uptime"[*] prevents the stop and go typical of non-LM operations.[†] Where TPM ends and special-cause problems arise, *employee involvement* employs the firsthand knowledge of the intricacies and complexities of the manufacturing operations, equipment, and practices embodied in the hands-on experience of the employees to more quickly and effectively solve problems.[‡] Errors in production scheduling are eliminated (or at least minimized) through the implementation of kanban production *pull* systems. With this, the correct mix of products, in the quantity required and at the time needed, is produced.[§] *Setup* time reduction contributes to the correct production mix and schedule by enabling smaller lot size manufacturing, increasing the responsiveness to changing customer needs.[¶] Finally, the output of the LM operations is predicted with increased accuracy using *statistical process control* (SPC) methods.

Manufacturing systems around the world share common traits: equipment that works as expected, orders that are completed when asked, production that does not make scrap, and so on. While this book acknowledges that the sociocultural side of manufacturing matters, we cannot escape the fact that the technical side of manufacturing is highly complemented by the elements of Lean.

The integration and observable outcomes of these LM elements then derive the first hypothesis:

> **Hypothesis 1**: Lean manufacturing practices are positively related to operating performance.

Moderating Influence of Culture on H1

The effectiveness of LM elements is tempered by the culture of the country where they are deployed. In an effort to understand the extent to which the

[*] JKS-W, 1969.

[†] S. Tsuchiya, *Quality Maintenance: Zero Defects through Equipment Management*, Productivity Press, Cambridge, MA (1992).

[‡] W. Niepce and E. Molleman, Characteristics of work organization in Lean production and sociotechnical system, *International Journal of Operations and Production Management* 16, 77–90 (1996).

[§] R.J. Schonberger, *Japanese Manufacturing Techniques: Nine Hidden Lessons in Simplicity*, Free Press, New York (1982); Schultz et al., Modeling and worker motivation (1998).

[¶] White, Pearson, and Wilson, JIT manufacturing (1999); Koufteros and Vonderembse, The impact of organizational structure (1998).

Lean elements are tempered, it is necessary to evaluate the degree of congruence between the characteristics that are the basis for GLOBE (Global Leadership and Organizational Behavior Effectiveness)* cultural dimensions and the underlying approach as shown in Table 3.1. Where high congruence exists between the GLOBE dimensions and the LM values, there is a positive cultural influence on the effectiveness of LM techniques. In contrast, low congruence shows a negative influence on LM effectiveness. In assessing these connections between the GLOBE dimensions and the LM values, a key aid is the theory of "The Corruption of Managerial Techniques by Organizations" by Daniel Lozeau et al.† This theory helps explain why the effectiveness congruence influences occur. Lozeau's theory describes that people can co-opt managerial methods. In other words, employees can apply management techniques in ways that serve their own purposes, beliefs, and values. Sometimes, this can mean strengthening a technique (e.g., a detail-oriented person using SPC) or weakening a technique (e.g., an autocratic person running a brainstorming meeting). What matters then is whether the managerial method is "congruent" (i.e., in agreement with) with the cultural beliefs and values of a workforce.

Table 3.1 shows the degree of influence and connectivity between each of the eight LM values and the associated GLOBE cultural dimensions. The key features of the table are to demonstrate the various attributes that underlie the GLOBE cultural dimension, the degree to which these attributes are related to the LM values, and the overall expected influence of the cultural dimension to the effectiveness of LM. The table depicts gray-highlighted regions where the intersection of the LM value and the cultural dimensions have a positive relationship to LM effectiveness; black shows a negative relationship on influence; and white indicates no influence or unrelated characteristics. The combination or composite effect of the interrelationship of these elements, as evaluated against Lozeau's theory, provides a picture of the overall impact on LM effectiveness (as shown on the bottom row of the chart). Each directional influence follows from

* "GLOBE is the acronym for 'Global Leadership and Organizational Behavior Effectiveness.'" "Conceived in 1991 by Robert J. House of the Wharton School of the University of Pennsylvania, and led by Professor House, the GLOBE Project directly involved 170 'country co-investigators' based in 62 of the world's cultures as well as a 14-member group of coordinators and research associates. This international team collected data from 17,300 middle managers in 951 organizations." From http://www.grovewell.com/pub-GLOBE-intro.html, accessed March 17, 2012.

† D. Lozeau, A. Langley, and J.-L. Denis, The corruption of managerial techniques by organizations, *Human Relations* 55 (5), 537–564 (2002).

TABLE 3.1

Development of Hypotheses

Lean Manufacturing Values	Institutional Collectivism (IC)	Uncertainty Avoidance (UA)	Assertiveness (AS)	Future Orientation (FO)	Human Orientation (HO)	Power Distance (PD)	In-group Collectivism (GC)	Gender Egalitarianism (GE)	Performance Orientation (PO)
L1. Base decisions on visible, first-hand facts and problems in a world that's difficult to predict	Relationships should preside over rationality and self-interest	Prefer to see world as predictable; laws and formal systems reduce ambiguity	Rational thought and direct communicaions are neeed	Plans can be long-term because the relevant information is known	Decisions are idiosyncratic, not standard; trust informal processes	Superiors know best; information is localized and discussion is unhelpful	Decisions based on "general will" rather than facts and rationality; indirect communication	Stereotypes should not interfere with decision making	Most situations can be controlled and predicted; results-based rewards and feedback needed
L2. Urgent timeliness toward production and quick improvements guided by long-term ideals	Long-term relationships are more important than short-term transactions, sacrifice individual interests for achieving collective goals	Long term future hard to predict, timely fulfilling immediate targets more important		Long time horizon, opportunities will come, less urgency to immediate signals		Power is stable over time and difficult to change in short-term	The pace of life should be slower and less stressful		Immediacy for results and a sense of urgency

continued

TABLE 3.1 (continued)

Development of Hypotheses

Lean Manufacturing Values	Institutional Collectivism (IC)	Uncertainty Avoidance (UA)	Assertiveness (AS)	Future Orientation (FO)	Human Orientation (HO)	Power Distance (PD)	In-group Collectivism (GC)	Gender Egalitarianism (GE)	Performance Orientation (PO)
L3. Problems are systemic and help reveal issues that people intrinsically are challenged to solve	People desire to achieve officially established group goals	Seek to reduce what's unknown; people are unpredictable and should conform	Personal responsibility for results in a controllable world; work hard toward difficult targets	Long-term outcomes matter most, immediate needs unclear	Altruism, benevolence, kindness, need for affiliation and self-sacrifice	Best to please and emulate superiors as they help maintain control and order	Conform to local "group will" and devalue intergroup influence	No gender-task misfit exists because problems lie in the process; capability not gender specific	Performance-based rewards needed to motivate employees in their personal desire for success
L4. Continuous, stepwise improvements prevent entropy, assure stability, and focus on learning by doing	Respect structures in place as they exist for a reason	Take moderate risks; use legitimate procedures and codes to keep stability	Personal judgment, habits, and preferences are important; rapid progress inspired by individuals is attractive	Knowledge acquisition helps future development; unexpected occurrences are just distracting		Concentrated power assures stability and order; new skills are unnecessary	Strive to maintain harmony and tradition	Skills not genetically determined but are improved by training, a greater acceptance of change	Continuous self-improvement and self-initiative; what you do is valued
L5. Productivity needs continuous monitoring with distinct work duties among operators and leaders	Teamwork and consensus are important; relation-based hiring wins over skill-based; training is always required	Need rules and process controls to increase trust and understand work	Environment can be changed to satisfy personal needs; competitive and opportunistic behavior are expected	Ideals can be attained, social relationships matter due to long-term working relationships	Social relations can be profitable; humans have rights and errors should be forgiven; resources should be needs-based	Titles and ranking are expected; automation is acceptable; personal choice is unimportant	Maintaining good within-group relations; duties and obligations direct behavior	People can decide right from wrong on their own and don't need institutions to assist them	Individual excellence, hard work, perseverance, and goal-attainment satisfy a personal need for achievement

L6. Entire workforce is interdependent and must cooperate using guidelines to solve problems	Resolve conflict with compromise and cooperaiton across groups; the collective concern is important	Group-processes provide assurances and avoid risks	Dominance and toughness are respected; people must trust capabilities, not obligations; cooperation not useful		A relationship orientation allows sensitivity to others; share power	Actions driven by commands from supervisors, not by needs to cooperate with co-workers; little cooperation across hierarchical levels	People are interdependent, but intragroup needs take precedent over intergroup productivity	A lesser role for guidelines, do not rely on others in decision-making, active and vocal members	Value independence and personal skills; people have a personal responsibility for successes, competition over cooperation
L7. Detailed routines create standards that temper irrationality and uncertainty while channeling human capabilities	Reward groups for performance; group decision-making is preferred, stick to group routines	Systems need control and planning; experts are helpful; standard procedures reduce variance	The self-needs of people direct behavior, not external controls or rules	Actions always influence the future; strategies should be made and current structures must be flexible toward goals	Work with others through mentoring; paternalism is valued; informal control of coworkers is expected	Workers aren't responsible and superiors make decisions, resources unequally distributed	Boundaries and strict lines of authority detract form cooperation; tolerate individual errors	Standardized procedures help reduce biases; gender-task fit doesn't exist	Responsibe to high standards with some uncertainty of success
L8. Self-benchmark to improve service performance using internal leader guidance	Organizations encourage employee to strive toward collective goals	External highly uncertain; be careful when selecting relations	Mindsets are internally focused; external needs unimportant		Internal means of solutions, informal relationships crucial	Higher status organizations and institutions should direct company	Distinguish between in- and out-group focus internally		External environment can be controlled; improving internal productivity; performance-based rewards

Source: W.P. Carey School of Business, Arizona State University.

the theoretical congruence between the worldview associated with a GLOBE dimension and LM's underlying mindset.

HYPOTHESIS 2: INSTITUTIONAL COLLECTIVISM POSITIVELY AFFECTS LM EFFECTIVENESS

As with culturally moderating influences we described previously, within the culture-Lean interaction there are other influences that provide a strong positive influence on the effectiveness of Lean efforts within an organization. Each of these influences possesses its own character and results in varying levels of positive contribution.

The first of these influences is *institutional collectivism* (IC). This dimension essentially describes the way in which an organization encourages, practices, and rewards "collective action" in contrast to individual action and performance.[*] IC influence within a culture reflects the degree to which the culture orders the priorities of group loyalty over individual goals, collective interests over individual gains, and group acceptance over individual achievements. The view that tends to drive IC in cultures and companies is the belief (or worldview) that because of the interdependence between people to one another and between people and the environment, where duties and obligations are crucial, harmony among these parts in the natural and social worlds should be the preferred choice of action as opposed to mastery of those worlds.[†] Second, the view of harmony also underpins the LM mentality regarding change and change management. In LM, change typically is approached in an incremental[‡] fashion to minimize the occurrence of unexpected disruptions. This also is culturally congruent in high-IC cultures. Third, high-IC cultures tend to more readily and rapidly accept the LM "system ideals" because these ideals refer to to the "collective whole" of the facility, once again emphasizing the group over the individual (or the whole over the parts). Finally, cultures that exhibit high IC will embed Lean practices more permanently

[*] House et al., *Culture, Leadership, and Organizations*, Sage, Newbury Park, CA (2004).

[†] S. Schwartz, Beyond individualism-collectivism: New cultural dimensions of values, in U. Kim, H.C. Triandis, C. Kagitcibasi, S.-C. Choi, and G. Yoon, Eds., *Individualism and Collectivism Theory Method and Applications*, pp. 85–122, Sage, Newbury Park, CA (1994).

[‡] Incremental change is often described in term of *kaizen* (both flow and point kaizen), although radical change *kaikaku* is employed where and when necessary.

into regular work routines and daily decision-making practices. Those cultures that have high IC typically demonstrate cultural values in which group empowerment takes priority over individual empowerment, and collective *inter*dependence is respected as a virtue.* There is high prevalence of and expectation of "parental-acting" organizations in both public and private organizations (governmental agencies and private companies).

Within cultures and companies possessing high IC, positive influence on LM also exhibits institutional resistance to changes in group duties and responsibilities, in existing organizational and work structures, and in institutional norms. These high-IC influences are, in many cases, in stark contrast to the low-IC cultures (particularly in the West, especially the United States), which tend to reward individuals rather than groups; permit, or even encourage, individual preferences to outweigh group standards; and in companies where employees and employees have shorter-term relationships and affiliations to employers.†

The GLOBE study has identified high-IC countries to include Southern Asia, Latin America, and the Middle East; and low-IC regions include Eastern Europe, Nordic Europe, and Anglo countries (e.g., United States, England, Australia). Within these contexts, there are specific outcomes from LM practices that can be expected. Employees from regions where high IC is an ingrained part of their culture will find that the LM's emphasis on routine work and work standards and collaboration within the workplace are congruent with the values they possess and therefore experience less difficulty in the implementation of LM initiatives. In the LM environment, production, inventory, supply chain, and other "problems" tend to be seen as systematic. That is, when issues arise there is a dependence on standardized problem-solving methods. As a result, improving system routines and adhering to established work standards are important.

In cultures that exhibit a low-IC mindset, employees will disagree with LM's underlying approach and will likely corrupt the LM practices.‡ In low-IC cultures, employees will tend to seek out improvement opportunities and suggestions that result in personal benefit more than systemic benefit. Standard work will be perceived and treated as superficial in this

* M.J. Gelfand, D.P.S. Bhawuk, L.H. Nishii, and D.J. Bechtold, Individualism and collectivism, in R.J. House, P.W. Dorfman, and V. Gupta, Eds., *Culture, Leadership and Organizations*, pp. 438–513. Sage, London (2004).

† D. Rousseau, *Psychological Contracts in Organizations: Understanding Written and Unwritten Agreements*, Sage, Thousand Oaks, CA (1995).

‡ Lozeau, Langley, and Denis, The corruption of managerial techniques (2002).

environment, and employees will tend to create and utilize individualized "work-arounds" in performing their work tasks. The converse is true in high-IC cultures, where conforming to standards and collaborating in the workplace is elemental to the development and improvement of manufacturing and other work systems.

LM Action Example

A principal characteristic of LM continuous improvement is the optimization of continuous production flow (FLOW). The effectiveness and efficiency of the FLOW process are enabled by ongoing improvements in the factory layout and through the conformance to standardized procedures.* The implementation of FLOW strategies with factory floor modifications will be (may be) seen and deployed differently depending on the cultural backdrop from which they occur. High-IC cultures, those valuing group norms, will tend to see any difficulties in FLOW implementation/effectiveness as a systems or standardization issue. They will be less likely to blame any lack of success on individuals and accept any shortfall as structural. As a result, the work team will direct its attention to improving the work cell structures and better utilizing human capabilities. In this way and at the foundation of LM values is the understanding that group-based rewards emerge from cooperative work. In low-IC cultures, where personal benefits are pursued, negative facts are "deemphasized" to avoid individual penalty. This "clouding of the facts" creates an ambiguous work environment where mixed messages lead to muddied improvement opportunities and decisions.

In summary, because conforming to standards and collaborating for system improvement are considered key to LM success, a high level of IC is congruent with LM values and the following hypothesis is given:

> **Hypothesis 2**: The degree of *institutional collectivism* in a manufacturing facility's country culture positively moderates (increases) LM effectiveness.

* Hyer, Brown, and Zimmerman, A socio-technical systems approach (1999); C.A. Yauch, Moving Toward Cellular Manufacturing: The Impact of Organizational Culture for Small Businesses, University of Wisconsin–Madison (2000); and C.A. Yauch and H.J. Steudel, Cellular manufacturing for small businesses: Key cultural factors that impact the conversion process, *Journal of Operations Management* 20, 593–617 (2002).

HYPOTHESIS 3: UNCERTAINTY AVOIDANCE POSITIVELY AFFECTS LM EFFECTIVENESS

The term *uncertainty avoidance* (UA) describes the degree or extent that a culture relies on value-specific procedures and established norms in an effort to reduce the occurrence or likelihood of unpredictable events.[*] This cultural dimension generally reflects the aversion toward risk within the given culture. Cultures high on this dimension will assess situations more critically for risk factors and are likely to be threatened by those risks they perceive. In a similar fashion, cultures with high UA focus on situational and operational ambiguity and possess a high degree of concern regarding this ambiguity. As with the IC dimension, high-UA cultures also tend toward consensus or even unanimity and are intolerant of dissent.

As opposed to the potential discord in disharmony created in situations of uncertainty, high-UA cultures tend to seek a greater degree of orderliness in professional and personal situations. Consistency, structure, and formalized procedures are valued as a means of establishing a greater sense of order.[†] The desire for the sense of order and reduced risk of uncertainty will result in cultures that possess a high reliance on governmental laws and regulations to govern greater aspects of daily living. In the work environment (and often in private life) members of a high-UA society seek frequent short-term feedback. Incremental change, whether in professional or personal life, is common as a means to further reduce ambiguity and uncertainty.

By comparison, cultures with low-UA characteristics will be more at ease with informal processes and less dependent or concerned with the level of orderliness sought in high-UA cultures. These cultures are often more tolerant of rule breaking and are somewhat more cavalier or less calculating when taking risks.[‡] Also in sharp contrast to high-UA cultures, those with low UA do not view dramatic change as life threatening. In fact, such change is often welcomed in low-UA cultures. The regions

---

[*] House et al., *Culture, Leadership, and Organizations* (2004).

[†] H.C. Triandis, The self and social-behavior in differing cultural contexts, *Psychological Review* 96, 506–520 (1989).

[‡] M. Sully de Luque and M. Javidan, Uncertainty avoidance, in R.J. House, P.J. Hanges, M. Javidan, P.W. Dorfman, and V. Gupta, Eds., *Culture, Leadership, and Organizations: The GLOBE Study of 62 Societies*, pp. 122–145. Sage, London. (2004)

of the world that demonstrate high UA (from the GLOBE study) include Southern Asia, Eastern Europe, Latin America, sub-Saharan Africa, and the Middle East; those with low UA include Anglo Europe, Germanic Europe, and Nordic Europe.

While high cultural UA may lead one to think that such cultures are often more simplistic with respect to perceiving what is right from what is wrong, such cultures actually possess an ability to perceive with fine resolution the complexities, interdependencies, and ambiguities in the world and in the workplace. This ability to see these characteristics creates a cultural environment that is harmonious with the essential principals and mentality of an LM system. Workers in a high-UA workplace tend to view broader horizon, long-term situations as unpredictable; they place high importance and priority on establishing and tracking to short-term targets. This perspective seems to be at odds with the generally accepted principle that in LM environments long-term objectives take priority over short-term actions. Nevertheless, this short-term focus can be observed in the responsiveness of employees to the frequent and often-subtle process signals that indicate existing or emerging problems in the LM system.

Within a high-UA LM work environment, workers possess a sense of urgency in addressing and analyzing operational signals. This urgency arises in the attempt to reduce ambiguity by acting on system signals prior to information "spoilage." By acting in the most immediate way possible, this sense of urgency prevents missing improvement opportunities. Preventing information spoilage and avoiding missed opportunities are essential for immediate understanding and learning in a complex and dynamic (ever-changing) LM production system.

The character of (excessive) risk avoidance in high-UA cultures is also consistent with the structured (incremental) improvement methods, including a "stepwise" approach, advocated by LM and Lean systems methodologies. Workers in such environments and cultures tend to be more at ease with system and work improvement efforts that not only are incremental but also follow a more strict and well-specified scientific method. Such methods are associated by high-UA employees as having a higher predictability of outcomes and are less ambiguous with respect to the causal factors that are part of the improvement being considered.

As a practical consideration, the propensity to avoid high levels of risk in high-UA cultures is essentially born out of the desire to reduce unpredictable variability in the LM process. As standard work provides detailed work routines and operating procedures, the use of and adherence to standard work in high-UA cultures is likely to be more disciplined. The minimization of variance of the manufacturing process creates the lowered risk and variability environment valued in high-UA cultures. The LM practices of rigorous and continuous process control, detailed standard work and work procedures, and strict conformance to quality and manufacturing performance standards are all oriented toward achieving increased predictability (reduced ambiguity) and therefore have great appeal in high-UA cultures.

These same LM values and practices become more challenging in cultures that have low (or lower) UA. Such low-UA cultures would tend to view the same controls, procedures, and performance standards as unnecessary and costly burdens that create a dampening effect on invention and innovation. While not rejecting LM approaches, cultures with low UA will "corrupt" the implementation of LM by deploying the practices only superficially while maintaining much of the prior methods. This can be seen in many ways in operational practice and in managerial practice. For example, on the shop floor a low-UA conversion to LM may include the development, posting, and implementation of standardized work procedures. However, as opposed to a disciplined adherence to the standard work procedures, low-UA cultures will allow the operators to vary their work as needed. Similarly, in Lean "leadership" managers will provide lip service to adhering to disciplined Lean practices while demanding the process be "short-circuited" if performance objectives are at risk (especially financial objectives). There tends to be a corruption of applying a disciplined approach to respond to measures of process as opposed to measures of results.* Managers and workers are very comfortable with doing things differently than standards, quite OK with processes not being highly detailed procedurally, very willing to take risks with major process changes without a thorough test process, and so on.

Superficial implementation of LM does not fully utilize LM's potential in improving system performance; therefore, the following hypothesis is given:

> **Hypothesis 3**: The degree of uncertainty avoidance in a manufacturing facility's country culture positively moderates Lean manufacturing effectiveness.

---

* David Mann, *Creating a Lean Culture: Tools to Sustain Lean Conversations*, 2nd ed., Productivity Press, New York, pp. 153–159 (2010).

HYPOTHESIS 4: ASSERTIVENESS NEGATIVELY MODERATES LM EFFECTIVENESS

While the characteristics of cultures exhibiting UA are typically associated with Eastern countries, cultural assertiveness (AS) is far more typical of Western cultures, especially countries such as the United States. The GLOBE study defined *assertiveness* as "the degree to which people value confidence, confrontational behavior, and the *aggressive* defense of one's position."[*] This cultural attribute might be popularly described as what Americans have come to know as "rugged individualism." High-AS individuals and cultures tend to have a "dominance" worldview, a world that is mechanistic and therefore controllable—one that can be dominated by human effort. Workers who see themselves in this mechanistic and controllable environment take full responsibility for their success and failures.[†] Assertive cultures exhibit individualism where a "doing" orientation and self-initiative are valued.[‡] The ability to manage and affect the manufacturing environment is seen as requiring the worker to possess personal knowledge. To obtain this knowledge, high-AS cultures value rational thought over emotion, a characteristic seemingly at odds with the typical understanding of the characteristics of assertive persons.

In contrast to valuing AS, low-AS cultures view confrontation and confrontational behavior as destructive. This is a striking comparison to high-AS cultures, in which confrontation is seen as a necessary ingredient for accomplishing performance goals. High-AS cultures, workers, and leadership teams will prioritize performance results over working relationships and foster competition over cooperation. These perspectives come from the view that assertive people are apt to see others as opportunists as they themselves tend to be opportunists. These other opportunists, because of their self-interest, are not seen as inherently trustworthy, as would likely be the case in low-AS cultures.[§]

---

[*] D.N.D. Hartog, Assertiveness, in R.J. House, P.J. Hanges, M. Javidan, P.W. Dorfman, and V. Gupta, *Culture, Leadership, and Organizations: The GLOBE Study of 62 Societies*, pp. 395–431. Sage, London (2004).

[†] F. Trompenaars and C. Hampden-Turner, *Riding the Waves of Culture: Understanding Cultural Diversity in Global Business*, 2nd ed., McGraw Hill, New York (1998).

[‡] F.R. Kluckhohn and F.L. Strodtbeck, *Variations in Value Orientations*, Greenwood Press, Westport, CT (1973).

[§] Regions of the world with high-AS values from the GLOBE study are Southern Asia and Confucian Asia; low-AS regions are the Middle East and Germanic Europe.

Of the various cultural values, cultural AS has one of the strongest correlations with the effectiveness of LM. The greater the cultural bias toward AS, the less effective LM is likely to be. This inverse effect is driven first by the emphasis of personal responsibility and knowledge and the prominence of results orientation in high-AS cultures. This perspective is in direct contradiction to the LM reliance on standardized work, where authorized manufacturing methods, corrective actions, and processes are used. High-AS workers rely far more on personal know-how in addressing work-related issues. Likewise, individual knowledge is valued in high-assertive cultures, whereas group knowledge is valued in low-AS cultures. Immediate learning is necessary in the LM environment, and as a collective activity, high-AS workers tend to be reluctant to share personal knowledge for the sake of collective knowledge.

In LM, the concept of continuous incremental improvements implies a never fully achievable ideal manufacturing system. Low-AS workers recognize this contradiction, yet are challenged (and motivated) by the improvement of the LM system and move toward continuous improvement. High-AS workers, in contrast, are motivated by performance targets that are achievable but often extremely challenging. If an ideal LM state or system is never fully achievable, there is not an attainable target; therefore, it is not motivating to a high-AS worker.

Problem identification and resolution also are inherent to an effective and efficient LM system and environment. It is the resolution of these problems that is the hallmark of an effective LM system. When problems are identified in high-AS cultures, the problem resolution is pointed toward the individual, not the systems. Production operators are blamed for system and process failures instead of the systems themselves. When this occurs, no real root cause analysis is completed, and the actual causes may be missed. People rarely are the actual root cause of system failure in the LM setting. Aggressive workers (and management teams) will tend to blame operators and workers in general for any manufacturing shortfall.

Finally, whether actually existing in an environment of crisis demanding expedited problem resolution or creating an artificial sense of urgency, high-AS cultures value and emphasize rapid progress. The nature of the rapid progress is driven by the individual personal judgment and preferences by the workers. This individualistic approach once again is contrary to LM values and practices, which emphasize structured and incremental improvement controlled by detailed routine.

When high-AS cultures implement LM systems, there is a high degree of the corruption of LM practices. This corruption often results in intrusive and destructive behaviors. Problem identification will tend not to serve as an opportunity for continued learning, but as a way to point blame toward an operator or other individual and force drastic improvements in an unprepared environment.* For example, SETUP (reduced setup times) requires an immediate response to any event causing an unexpectedly long setup time.† High-AS cultures will tend to blame the operator for not working hard enough, instead of tending to blame the system; opportunities for removing root causes will not be exploited, and longer setup times will result.

Cultures high in AS place great emphasis on personal responsibility, individual judgment, and self-needs and demonstrate propensity toward rapid and radical progress. These characteristics are not congruent with the foundational principals of LM. Therefore, the following hypothesis is given:

> **Hypothesis 4**: The degree of assertiveness in the culture of the country of a manufacturing facility negatively moderates LM effectiveness.

HYPOTHESIS 5: FUTURE ORIENTATION NEGATIVELY MODERATES LM EFFECTIVENESS

As mentioned in the opening chapters, the characteristic of future orientation (FO) with respect to LM effectiveness is often the one most misunderstood by Western practitioners of LM and other Lean system approaches. High-FO cultures will likely "stick to the plan" regardless of new information. This can mean they will be less adaptable. In a stable environment, this can make sense. But, in an LM plant in tune with changing customer demands, sticking to production plans is not going to work. High-FO cultures will not have the sense of urgency—they are often associated with the "cultures of mañana." High-FO cultures feel the

* While LM systems emphasize incremental planned improvements, typically through kaizen efforts (both point and flow kaizen), there are times when radical change (kaikaku) is appropriate. High-AS cultures will often minimize the value of incremental (kaizen) improvement and treat all improvement needs as drastic or radical (kaikaku).

† White, Pearson, and Wilson, JIT manufacturing (1999).

future today. That is, if something will get done tomorrow, that is almost as good as it getting done today. If it eventually will get done, then that counts for now.

As it is defined in the GLOBE study, FO is the degree to which a society, and the individual members of that society, see the context of their present actions as affecting or influencing the future state of things and, as a result, to what extent long-term planning should be used as a managing factor for the future state outcomes.* From high-FO countries, individuals tend to see the world beyond its present state. Being interested in, and perhaps primarily driven by, the future state, high-FO individuals are often uninterested in and less responsive to immediate issues.† In contrast to a view of the world as dynamic, interconnected, interdependent, and somewhat unpredictable, high-FO cultures perceive a mechanistic world of predictable cause-and-effect outcomes. Such cultures with this mechanistic worldview will believe that all the relevant information needed to affect future outcomes is readily available, and that the future can be reliably predicted from this information.

By contrast, low-FO cultures pay much greater attention to current conditions and are acutely aware of the effects of these conditions. The general worldview is that a long-term or long-time picture of the future is unpredictable, based on a less-mechanistic and more dynamic understanding of factors potentially affecting outcomes. As a result, workers in low-FO cultures are more responsive to deviations in current operating performance relative to immediate targets.‡

Most Western practitioners attribute an FO as a success factor in the deployment and sustainability of LM systems. Their anecdotal evidence to support this is to point to the long-term planning perspective of companies such as Sony, Toyota, and others from the East with respect to their market penetration and ultimate dominance perspective. Southern Asia is

---

* House et al., *Culture, Leadership, and Organizations* (2004).

† T.T.D. Peetsma, Future Time Perspective as an Attitude: The Validation of a Concept, the 5th Conference of the European Association for Research on Learning and Instruction (EARLI), Aix-en-Provence, France, 1993; M.J. Rosenberg and C.I. Hovland, Cognitive, affective, and behavioral components of attitudes, in C.I. Hovland and M.J. Rosenberg, Eds., *Attitude Organization and Change*, pp. 1–14, Yale University Press, New Haven, CT (1960); R. Seginer and R. Schlesinger, Adolescents' future orientation in time and place: The case of the Israeli kibbutz, *International Journal of Behavioral Development* 22, 151–167 (1998); G. Trommsdorff, Future orientation and socialization, *International Journal of Psychology* 18 (5), 381 (1983).

‡ K.A. Keough, P.G. Zimbardo, and J.N. Boyd, Who's Smoking, Drinking, and Using Drugs? Time Perspective as a Predictor of Substance Use, *Basic and Applied Social Psychology* 21, 149–164 (1999).

in fact among those countries that exhibit high-FO values.* However, high FO directly conflicts with many of LM's underlying values and mindset.

The conflict between high-FO values and LM values can be seen first in the orientation itself. High-FO cultures value and adhere to long-term plans because they believe that the world is highly predictable and readily "knowable." This view conflicts with LM values in a number of ways. An FO leads to a lessened sense of urgency in resolving deviation from current situations and short-term goals—directly conflicting with Lean value 2 (LM2 time). In addition, the view of reliable future predictability also defeats much of the purpose of standard work, which by its nature is meant to reduce operating uncertainty. LM practices and systems are built on a view of continuous improvement, one in which no perfect system ideal is obtained, yet is still strived for. High-FO cultures optimistically view that such perfected ideals are attainable.

High-FO cultures are also likely to ignore the visual signals and improvement opportunities that arise in short-term performance. LM's postulation that the world is hard to predict requires workers to pay firm attention to and grasp current conditions, conditions to which immediate signals require special attention (LM1 truth and LM3 motivation). High-FO cultures will look beyond the short-term need for the long-term plan. Closely associated with this is the LM view of incremental improvements, where after making incremental change the effects of the change are observed to understand their effect. This evaluative step conflicts with the FO perspective of planning many steps into the future. FO workers will default to the FO versus responding to what is currently needed.

As with other negatively moderating cultural elements, FO results in the corruption of LM practices. These practices are corrupted due to an ignorance of the immediate signals and clues to cause-and-effect relationships, thus missing many opportunities for system improvements. While facilities in high-FO cultures may welcome the plan of using LM as a way to improve their future prospects, they will miss most of what LM is meant to do. Corruption will occur. Lengthy plans, extensive meetings, and deep analyses are likely to be connected to every improvement activity. Overall, a culture valuing high FO will find it difficult to fully realize LM's potential in improving system productivity. Therefore, the following is proposed:

* Regions of the world with high-FO values from the GLOBE study are Southern Asia, Latin America, Middle East, sub-Saharan Africa; low-FO regions are Nordic Europe and Germanic Europe.

Hypothesis 5: The degree of future orientation in the culture of the country of a manufacturing facility negatively moderates LM effectiveness.

HYPOTHESIS 6: PERFORMANCE ORIENTATION NEGATIVELY MODERATES LM EFFECTIVENESS

Performance orientation (PO) has a similar negative moderation on LM effectiveness as FO. It also has a similar misunderstanding by practitioners in that it misperceives performance in LM—confusing performance measures of process versus performance of outcomes. In Western cultures, performance of outcomes typically outweighs performance of process—including it being the basis for compensation and performance awards.

PO is the value by which cultures collectively encourage members to excel among their peers. Also characteristic of PO cultures is that their members have an internal "locus of control" in relation to the world versus an external locus of control.[*] This means that high-PO cultures do not view people as dependent on the environment, but people as able to influence and control the environment to their own ends. This is another perspective of the world as mechanistic, subject to cause and effect of personal world shaping. In this mechanistic worldview, this cultural aspect lends rise to demonstrating the degree of control exhibited by individuals—that is, a need for achievement.[†] Workers in high-PO cultures also exhibit a need to find importance in one's work; to place great emphasis on ambitious expectation for self, career, and performance; and to demonstrate responsibility for success, independence, and growth of personal skills.[‡]

Because of a dependence on achievement, specificity in performance is required for workers in high-PO cultures. These employees respond better in work environments that provide concrete targets versus general performance principles. Employees in low-PO cultures prefer more holistic-relational approaches to their work and careers. This cultural value is somewhat mediated, however, with the majority of GLOBE regions clustered toward a "medium" degree of PO values. There are high-PO cultures in Latin America and low-PO cultures in Confucian Asia.

---

[*] Trompenaars and Hampden-Turner, *Riding the Waves* (1998).

[†] D.C. McClelland, *Human Motivation*, Cambridge University Press, Cambridge (1987).

[‡] M. Javidan, Performance orientation, in R.J. House, P.J. Hanges, M. Javidan, P.W. Dorfman, and V. Gupta, *Culture, Leadership, and Organizations: The GLOBE Study of 62 Societies*, pp. 239–276. Sage, London (2004).

The negative moderation of PO cultures stems from the emphasis on personal skills over systems knowledge in high-PO cultures. Because of the LM focus on improvement and better systems over personal development, immediate learning in response to systems problems will not be motivating to high-PO employees. Although employees in high-PO cultures are motivated by learning when solving problems, their focus is on individual knowledge (for the purpose of achievement) versus shared system knowledge. As a result, high-PO employees are likely to corrupt LM practices in a number of ways. First, these employees are less likely to drive as deep in determining root cause, thereby understating or subverting a full understanding of the production complexities giving rise to operating problems. Second, because high-PO employees are individually (internally) focused with respect to performance and success, they are also likely to ignore or suppress any facts that they may believe would cast a bad light on their performance. This means that problem identification and exposure are less likely to happen, contrary to LM values and practices. Also, with a greater level of peer competition, worker collaboration will suffer, and systems problems will not be cooperatively identified. In this environment, LM will only be superficial; only obvious system problems are identified and other issues covered up—including those "minor" issues that may be more useful in understanding system performance.*

The influence of peer competition is important to understand. This is key—high-PO cultures are very concerned about peers. They are satisfied when they are ahead of everyone else, not as satisfied as when their peers are ahead of them. Self-improvement is important, but once an employee exceeds others, then the motivation for improvement diminishes. And again, the emphasis is on self-improvement, not on system improvement.

As with FO, workers in PO cultures respond better to very specific performance goals based on an improved or ideal future state; that is, they respond better to an *attainable* future state—not some general principle or ideal. Because an ideal future LM state does not exist, high-PO employees will be uncomfortable in a pure LM environment. High-PO workers will feel "directionless" in an LM system because expectations of performance are generally determined by trial and error as a result of immediate attention to systems issues and problem solving. Finally, in high-PO work

* K. Suzaki, *The New Manufacturing Challenge: Techniques for Continuous Improvement*, Free Press, New York (1987).

cultures, there is great emphasis on individual excellence, independence, and competition. This environment makes the use of standard work* difficult because these characteristics disassociate the outcomes from individual achievement and outcome. With standard work, a successful (or unsuccessful) outcome is directly attributed to the process, adherence to the process, and not the individual. For example, performance such as zero defects is seen as a reflection of good standards in LM, not because of the individual effort of skilled operators. Because LM inhibits the employees' ability to differentiate themselves and their performance from their peers, employees in high-PO cultures are likely to corrupt the LM system by personalizing their work. In this way, they are able to claim that extraordinary performance is an indicator of personal excellence versus system excellence. Therefore, the following is proposed:

> **Hypothesis 6**: The degree of performance orientation in the culture of the country of a manufacturing facility negatively moderates LM effectiveness.

DIMENSIONS WITH NO NET INFLUENCE ON LM EFFECTIVENESS

Having already covered the GLOBE cultural dimensions that have a positive effect on LM values and those that have a negative effect, we briefly examine those that have no *net* effect on LM effectiveness. Another way to think about these dimensions is to understand them as culturally neutral to LM or having no consistently positive or negative impact.

There are four GLOBE cultural dimensions in this group: gender egalitarianism (GE), in-group collectivism (GC), power distance (PD), and humane orientation (HO). As shown in Table 3.1 previously, these dimensions show some influence, even sometimes significant influence, but there are a confounding relationship and effects of these characteristics. While no direct conclusions could be drawn regarding the effect of these four attributes, we include them here to provide an understanding of their existence and possible effect on LM effectiveness and for completeness of the research and analysis.

* This would include both manufacturing/assembly standard work and standard work for leaders.

Gender egalitarianism is an analysis of the degree with which a society minimizes gender inequality.[*] While not precisely the same, this dimension can be stated in the positive: GE is the measure of the degree of gender equality within a culture. As would be anticipated, countries with low-GE cultures believe in the genetic determination of skills and attributes. The behaviors expected and exhibited by the various genders are determined by gender, and these behavioral stereotypes are in the best interest of society and should be retained.[†]

In contrast to the gender stereotypic view of low-GE societies, countries with high GE do not believe in gender predetermination. These cultures tend to demonstrate a higher level of political activism, are typically more vocal in expressing options and ideas, and are more receptive to change.[‡] An example of the "confounding effect" of this dimension can be seen in high-GE cultures where the world is seen as ever changing and requiring adaptation. These cultures recognize the benefit of and need for progress, so welcome the LM practice of problem identification and root cause analysis, for which striving for a system ideal will be welcome. Yet, these same societies also have a disregard for strict roles. Their members believe that individuals decide what is "right" versus the decisions being made by institutions. Here, the LM practice of standard work and structured improvement tend to be corrupted. In a sense, GE can simultaneously have a positive and negative moderating effect on LM effectiveness—or overall, have no net effect.

In-group collectivism is a measure of the degree to which a society values loyalty, pride, and cohesiveness. Cultures that emphasize this value among their people are high-GC cultures.[§] The GC dimension reflects the security individuals assume or feel within society. This security is based on the perception of the individuals' participation in, or identity with, a work team, family, or other tight-knit group and the "power" that is bestowed on them as a result of group membership. These groups form norms of reciprocity, solidify identity and affiliation with these "tribal" organizations, and influence members' behaviors toward the best interest of the group.

[*] House et al., *Culture, Leadership, and Organizations* (2004).

[†] G. Hofstede, *Culture's Consequences: Comparing Values, Behaviors, Institutions and Organizations across Nations*, 2nd ed., Sage, Thousand Oaks, CA (2001); D.L. Best, J.E. Williams, J.M. Cloud, S.W. Davis, L.S. Robertson, J.R. Edwards, H. Giles, and J. Fowles, Development of sex-trait stereotypes among young children in the United States, England, and Ireland, *Child Development* 48, 1375–1384 (1977).

[‡] House et al., *Culture, Leadership, and Organizations* (2004).

[§] Gelfand et al., Individualism and collectivism (2004).

The positive moderating influence of high-GC cultures is seen in the way in which the members see the world and their place in it. Cultures high in GC possess a worldview that is interpersonal and interdependent and in which maintaining harmony is a valued trait. These characteristics are congruent with LM's emphasis on collaboration within the context of immediate learning and structured improvement. These same cultures, however, are highly affected by "group will"; a collective mindset drives action and behavior. This trait will be in direct conflict with the LM practice of letting problem exposure drive improvement. Likewise, the high work team affiliation will result in a lessened sense of cooperation *between* work teams than the priority of work within the team. This action impedes the effectiveness of standard work that seeks to balance work between cells. Because of these counteracting effects, no relationship is hypothesized between GC and LM effectiveness.

Power distance is another of the GLOBE characteristics that tends to be misunderstood and misapplied by practitioners in their application of LM. This characteristic reflects the degree to which members of a culture expect organizational power to be concentrated at higher levels of an organization.[*] This organizational model is often associated with Japanese companies, such as Toyota, the typical model of LM success. Such companies possess a high level of employee participation, yet the GLOBE dimension suggests that high PD dampens disagreements with organizational superiors. This reluctance to express disagreement with superiors promotes autocratic decision making over democratic-like organizations.[†] In contrast, low-PD cultures expect that independent operational decision making will be made by individuals irrespective of the decision makers' hierarchical level.[‡] In a low-PD culture, employees at lower hierarchical levels are more likely to contribute to decision making through communicating problems and decisions to their superiors.

Production facilities in countries and cultures with high-PD values may find it difficult to have effective LM practices. High-PD cultures value hierarchy and hierarchical controls as a method to increase production stability and operational predictability.[§] While this is consistent with LM's

[*] House et al., *Culture, Leadership, and Organizations* (2004).

[†] Hofstede, *Culture's Consequences* (2001).

[‡] C. Nakata and K. Sivakumar, National culture and new product development: An integrative review, *The Journal of Marketing* 60 (1), 61–72 (1996).

[§] D. Carl, V. Gupta, and M. Javidan, Power distance, in R.J. House, P.J. Hanges, M. Javidan, P.W. Dorfman, and V. Gupta, Eds., *Leadership, Culture and Organization: The Globe Study of 62 Societies*, pp. 512–563, Sage, Thousand Oaks, CA (2004).

emphasis on standardization and structured activities geared toward complexity reduction, high-PD cultures discourage lower-level employees from contributing to system design. There is also a lack of cooperation between hierarchical levels, similar to the intrateam cooperation issue discussed previously. These factors conspire to create a corrupted LM environment in which production operators are either unable or unwilling to reveal production problems or share tacit process knowledge, resulting in inferior process and cell knowledge.[*] Due to these counteracting effects, we do not hypothesize a significant relationship between PD and LM effectiveness.

Humane orientation is the measure of a culture's encouragement and reward to its members for being fair, altruistic, friendly, generous, caring, and kind.[†] This cultural value reflects the view of the members of a society toward well-being of others. In high-HO cultures, welfare, informal patronage relations, and paternalistic norms outweigh achievement, rationality, formal procedures, and rewards.[‡] People in high-HO cultures believe others are important and are urged to promote their well-being. In such societies, people are motivated by the need for belonging and affiliation instead of power and material possessions. Behaviors are more controlled by paternalistic norms and informal (patronage) relationships than by formal procedures and rules. In high-HO cultures, welfare, relations, and needs outweigh achievement, rationality, and rewards.

There are elements of humane orientation that are consistent with the underlying mindset of LM. Good social relationships characteristic of HO cultures contribute to cooperation when exposing problems and seeking immediate learning. Also, high-HO cultures encourage altruism, a characteristic that facilitates system-oriented learning rather than self-focused learning. However, high-HO cultures also see work primarily as a means to have "a comfortable life."[§] This perspective is incongruent with the LM focus on developing a system ideal. High-HO cultures will not embrace the structured improvements or standard work systems that may make work life more stressful or one that runs counter to interpersonal advice. Based

---

[*] Hyer, Brown, and Zimmerman, *A socio-technical systems approach*, pp. 179–203 (1999).

[†] House et al., *Culture, Leadership, and Organizations* (2004).

[‡] H. Kabasakal and M. Bodur, Humane orientation in societies, organizations, and leader attributes, in R.J. House, P.J. Hanges, M. Javidan, P.W. Dorfman, and V. Gupta, *Culture, Leadership, and Organizations: The GLOBE Study of 62 Societies*, Sage, London (2004).

[§] Ibid.

on the counteracting effects, we do not hypothesize that HO significantly influences LM effectiveness.

SUMMARY

From the GLOBE study of cultural values, there are a number of culture-specific characteristics that affect the effective deployment and sustainability of LM efforts. There are a number of ways in which these cultural values intersect LM. Those that positively influence LM effectiveness include IC and UA. There are others that have a negative moderation on LM effectiveness. The negatively moderating values are AS, FO, and PO. Finally, there are a number of GLOBE cultural dimensions that have no net influence on LM, although that is not to say that there is no effect. The dimension with no net influence may actually positively and negatively affect LM at the same time, resulting in a neutral net impact. These cultural dimensions with no net influence are GE, GC, PD, and HO.

The no-effect dimensions often seem the hardest to place in the context of cultural influence. A better way to understand these might be to recognize that just because a culture is different does not mean it will have an effect on LM effectiveness. Many cultures are highly different on such things as GE and PD. In particular, Japan has a high PD and a low GE. Practitioners and academics alike make the mistake in thinking that Lean works in Japanese-type cultures. Based on the hypotheses and theoretical underpinnings, that will not be the case. We can note what Rother and others have mentioned, that *Toyota's organizational culture is not Japanese culture.* Those at Toyota intentionally separate themselves from mainstream Japan, keeping a "country bumpkin" persona. The practical insight is this: Cultural differences do not automatically equate to effect differences. What matters are the specific culture trait/dimensional congruence and incongruence and how to prepare an organization or adapt a managerial practice to attain more congruence.

It is the first hypothesis, however, that underpins the effect of all the GLOBE cultural dimensions. Perhaps a statement of the obvious, this first hypothesis is that *Lean manufacturing practices are positively related to operating performance.* It is necessary to establish this hypothesis in an effort to differentiate the cultural aspect of LM within the character of an organization in contrast to other productivity or quality improvement initiatives. While many other à la carte efforts have the potential to contribute to isolated improvements, these efforts do not necessarily provide sustained improvement through immediate action and learning, where an ideal future state is pursued but never attained. It is for this reason that you see other well-known improvement efforts, such as 6σ (Six Sigma), adopt or incorporate "Lean" as part of their effort or name. In many applications, for example, 6σ has become "Lean 6σ." The desire to capture the cultural elements beyond the limited scope of the "improvement" work and into the overall character of the company and its management is what these repackaging and remarketing efforts are about.

QUESTIONS TO CONSIDER

1. Are you familiar with the global cultural dimension of values that may affect a successful practice of LM?
2. How do your Lean practices need to be modified in various cultural applications to counteract any negative effects of individual cultural values—without compromising the core purpose of LM and systems?
3. Do you know which cultural values may have a positive effect on LM practices, and how you could maximize those influences?
4. Are your LM practices being "corrupted" to conform to a cultural value? If so, what mitigation efforts do you exercise to correct this corruption?
5. Have you or your company developed a multicultural LM implementation plan that accounts for variations in cultural values and practices? That is, how do you make Lean systems and efforts transferrable throughout the world?
6. Are your customers asking your company to implement practices that they call "Lean" but that they have "corrupted"? If so, how do you respond?
7. Considering the internal accommodations that must be made in addressing global cultural influence on LM, how does your organization apply these learnings to select and develop suppliers in different cultures?

4

Measuring Lean Culture

METHODOLOGY

Understanding the methodology of how Dr. Kull and his team developed the working basis from which to test the hypotheses outlined in Chapter 3 may only appeal to the academics and students who use this book as a course text or research source; however, we would encourage the practitioner not to bypass this chapter. Even if you have to return to the research methods after distilling the practical applications described, be sure to return as the methodology reflects the data-driven approach in Lean manufacturing (LM) that drives system change and improvement, as well as the way in which quality systems are developed and part quality verified.

Considering the broad range of the hypotheses we presented that proposes the way that global culture affects LM, testing of the hypotheses requires that the number of countries/global cultures and manufacturing facilities within those countries be examined.* As we have indicated throughout the preceding chapters, the global and cultural analysis was derived from the data developed by the Global Manufacturing Research Group (GMRG). Specifically, the hypotheses were tested against the data from the fourth-round GMRG worldwide survey. The data sets of the GMRG studies include manufacturing facilities around the world and consist of academic researchers from over twenty countries. The studies of these manufacturing sites were designed to assess and improve

* Because changes in standard errors are inversely proportional to the square root of the number of countries (Snijders and Bosker, 1999), a minimum of 13 countries would yield statistically significant *t* values, assuming typical effect sizes and standard errors of previous studies (Kull and Wacker, 2010).

manufacturing practices worldwide.[*] To provide the necessary cultural relevance, the GMRG survey questionnaire was translated and then back-translated by academics from each country, with the language occasionally modified to appropriately fit the local culture. This survey method/instrument has been used since 1985.[†]

As the research basis to validate the proposed cultural influences on the effectiveness of LM, the full GMRG 4.0 database was used. This database contains over 1,450 samples of manufacturing sites from twenty-four countries and includes twenty-two manufacturing industry classifications. Within this data set, the manufacturing plants represent both large and small operations. The majority of the observations came from plants with 50–500 employees. Of the total population, 14.1% were from manufacturing locations with over 500 employees (see the Appendix). Based on the overall sample size, the country and industry classification scope, and the range of manufacturing plant employee population, the GMRG data provide a good foundation on which to validate the effectiveness of LM across global cultures.

Within the GMRG data, there are survey questions that relate to operational performance and that indicate the presence of LM. These narrower data sets are those used for testing the LM effectiveness hypotheses. Particularly, questions were examined in which each manufacturing facility reported items that related to cost, quality, and delivery performance relative to its competitors (see Appendix Table A2). We also noted in these data that, on average, the respondents indicated that their competition, and therefore those they benchmarked themselves against, were domestic competitors.[‡] Also specifically examined were the reported responses to the way in which resources (money, time, or people) were invested toward LM-related programs.

---

[*] D.C. Whybark, GMRG survey research in operations management, *International Journal of Operations and Production Management* 17 (7–8), 686–696 (1997).

[†] C. Whybark, J. Wacker, and C. Sheu, The evolution of an international academic manufacturing survey, *Decision Line* 40, 17–19 (2009).

[‡] This method of analyzing the performance effects have been used by others, for example, K.O. Cua, K.E. McKone, and R.G. Schroeder, Relationships between implementation of TQM, JIT, and TPM and manufacturing performance, *Journal of Operations Management* 19 (6), 675–694 (2001); B.B. Flynn, S. Sakakibara, and R.G. Schroeder, Relationship between JIT and TQM: Practices and performance, *Academy of Management Journal* 38 (5), 1325 (1995); McKone et al., The impact of total productive maintenance (2001); M. Naor, K. Linderman, and R.G. Schroeder, The globalization of operations in Eastern and Western Countries: Unpacking the relationship between national and organizational culture and its impact on manufacturing performance, *Journal of Operations Management* 28 (3), 194–205 (2010); Shah and Ward, Lean manufacturing (2003).

The data captured and specifically examined the "core" LM components we have been discussing throughout the book. It was approached in this fashion remembering that LM is a subset of Lean production, with Lean production also including supplier-related practices.[*] The core LM components targeted were attaining continuous FLOW; having total productive maintenance (TPM); allowing and encouraging employee involvement (EMP); implementing the PULL production system; reducing SETUP times; and using statistical process control (SPC).

In the analysis of these components, seven LM applications were associated according to the following list: *cellular manufacturing*, associated with PULL and EMP[†]; time compression of *setup reduction* as reflected in the SETUP core attribute[‡]; *just-in-time* manufacturing and inventory placement derived from the applications of PULL, TPM, EMP, and SETUP[§]; *process redesign*, which occurs with TPM, SETUP, and FLOW[¶]; *throughput time reduction* reflects the application and effects of PULL and FLOW core attributes[**]; *statistical process control* disciplines stemming from SPC and that require EMP[††]; and *waste reduction*, which is the central theme of LM.[‡‡] The presence of these applications (or programs) as reflected in the data demonstrates the extent to which LM was present within the surveyed manufacturing facility. The GMRG survey method avoids survey bias or suggesting that the respondents link manufacturing responses to LM practices by separating questions of manufacturing performance and LM-related question on the survey itself.[§§] This separation was essential in the analysis to maintain a data separation of performance results to LM practices actually present within a facility.[¶¶] The

---

[*] Shah and Ward, Defining and developing (2007).

[†] V.L. Huber and K.A. Brown, Human resource issues in cellular manufacturing: A sociotechnical analysis, *Journal of Operations Management* 10 (1), 138–159 (1991).

[‡] S.C. Trovinger and R.E. Bohn, Setup time reduction for electronics assembly: Combining simple (SMED) and IT-based methods, *Production and Operations Management* 14 (2), 205 (2005).

[§] McLachlin, Management initiatives (1997); Sakakibara et al., The impact of just-in-time manufacturing (1997).

[¶] Cua, McKone, and Schroeder, Relationships (2001); Womack and Jones, *Lean Thinking* (1996).

[**] W.J. Hopp and M.L. Spearman, To pull or not to pull: What is the question? *Manufacturing and Service Operations Management* 6, 133 (2004).

[††] M. Rungtusanatham, Beyond improved quality: The motivational effects of statistical process control, *Journal of Operations Management* 19 (6), 653–673 (2001).

[‡‡] T.R. Browning and R.D. Health, Reconceptualizing the effects of Lean on production costs with evidence from the F-22 program, *Journal of Operations Management* 27 (1), 23–44 (2009).

[§§] P.M. Podsakoff, S.B. MacKenzie, J.Y. Lee, and N.P. Podsakoff, Common method biases in behavioral research: A critical review of the literature and recommended remedies, *Journal of Applied Psychology* 88, 879–903 (2003).

[¶¶] A one-factor analysis accounts for only 34.4% of data variance.

differentiation of practice monitoring and performance monitoring was highlighted in Chapter 3.

The GMRG survey results provided an abundance of data due to the extensive country, culture, and site population, coupled with the survey questions themselves. To identify the cultural effects of the LM state as an influence on or determinant of performance results, it was necessary to perform data reduction to simplify hypothesis testing *without* loss of theoretical insights. To accomplish this, one factor was extracted from the data for LM based on the data (for each hypothesis presented). First-order factors for cost, quality, and delivery were used from the data set to compute a second-order performance factor. The correlation of these factors provides an indicator of the influence the particular cultural element had on LM effectiveness.* In layman's terms, the results from the study of manufacturing effectiveness in various cultures were compared to the use/success of Lean in the same culture to show whether the culture positively impacted, negatively impacted, or had no effect on LM success in each of the LM core areas.

USING THE GLOBE MEASURES OF CULTURE

Global expansion of the manufacturing footprint that had previously been associated with first-world countries in both the East and the West has now grown at an exponential rate into emerging and third-world economies. Understanding of the cultural influences on market, manufacturing, labor, and event political situations has taken on a heightened sense of urgency. Much of inter- and multinational manufacturing experience has been the result of trial and error. Entry into new global manufacturing locations often has limited intelligence with respect to the practical launch

* Four factors are extracted by principal component analysis with varimax rotation using Kaiser normalization in SPSS 17.0. The factors are distinct, with eigenvalues of 4.8, 2.0, 1.3, and 1.1, accounting for 66.1% of data variance. The rotated component matrix exhibits acceptable item-to-factor correlations that are above the 0.5 threshold. Each factor exhibits high reliability; the Cronbach alphas for cost performance (a = 0.86), quality performance (a = 0.81), delivery performance (a = 0.90), and lean manufacturing practices (a = 0.80) are above the 0.70 threshold (Nunnally and Bernstein, 1994). Correlations of within-factor items are higher than nonfactor items, showing satisfactory unidimensionality. Mean item values for the three-factor operating performance (OP) dimensions are averaged, as are the seven item values for LM. Thus, single values for both OP and LM are estimated for each facility.

risks within the new geographic location—despite corporate claims of detailed due diligence. Most expansion relies heavily on local "tribal knowledge" and manufacturing adaptations to "make things work." Even though much more "systematized," LM is also subject to the cultural influences on the countries in which it is deployed. As a result, an understanding of culture is a critical *predictor* of operational success.

In the analysis, to measure culture, the results of the GLOBE study were used. These data are among the most exhaustive and are used widely to test cultural hypotheses.[*] The GLOBE research was complete in 1997 after ten years of studying culture in sixty-two countries, across 951 organizations, and surveying about 17,000 managers. This updated study expanded the original 1980 work to include nine cultural dimensions and other more detailed societal and leadership values. The GLOBE research included the development of measurement scales through a Q-sort technique, a technique that uses subjective viewpoints of those surveyed.[†] The survey method also relied on forward and backward translation as described previously to ensure a greater degree of accuracy regarding local language, customs, and the like. The data and data gathering included multiple pilot studies, multilevel factor analyses, and reliability and validity tests.[‡] The method described provides an especially useful database for the testing of global LM effectiveness as it provides "response-bias corrected scores" for each LM dimension for most countries surveyed. For those countries without LM effective scores, the cultural and population similarities of other surveyed countries provide sufficient and effective surrogate indicators.[§] For example, information for Ghana was estimated using West African scores (Hale and Fields, 2007). Macedonian information was estimated by averaging Greece and Albania scores (Poulton, 2000). Information

[*] M.C. Euwema, H. Wendt, and H. Van Emmerik, Leadership styles and group organizational citizenship behavior across cultures, *Journal of Organizational Behavior* 28, 1035–1057 (2007); and Y.D. Luo, Procedural fairness and interfirm cooperation in strategic alliances, *Strategic Management Journal* 29, 27–46 (2008).

[†] Steven Brown (1996) of Kent State University described the Q methodology (developed by British physicist-psychologist William Stephenson) as a method to identify and report the subjective observations or subjectivity of a given situation. The Q-sort technique is the basis of data gathering and analysis ranking agreement and disagreement of a statement.

[‡] See P.J. Hanges and M.W. Dickson, The development and validation of the GLOBE culture and leadership scales, in R.J. House, P.J. Hanges, M. Javidan, P.W. Dorfman, and V. Gupta, *Leadership, Culture, and Organizations: The GLOBE Study of 62 Societies*, pp. 122–145. Sage, London (2004).

[§] V. Gupta and P.J. Hanges, Regional and climate clustering of societal cultures, in R.J. House, P.J. Hanges, M. Javidan, P.W. Dorfman, and V. Gupta, *Culture, Leadership, and Organizations: The GLOBE Study of 62 Societies*, Sage, London (2004).

for Croatia was estimated by averaging the Macedonian estimate and Hungary scores (Tipuric, Podrug, and Hruska, 2007). Fiji information was estimated by averaging scores from the Philippines, Indonesia, and India (Norton, 1993).

HIERARCHICAL LINEAR MODEL APPROACH

The importance of the describing the data approach was not only to provide the basis by which the hypotheses were developed but also, more importantly, to indicate what managerial conclusions and actions can be drawn from the observations and how LM effectiveness can be shaped in various countries. That being said, this may be the point where practitioners may want to shield their eyes and jump ahead to Chapter 5. The following description of the hierarchical linear model (HLM) is important to LM process experts, quality control (QC) process owners, and academic researchers. We include it here to further develop the basis for reaching the results and managerial results.

The data validation of our hypothesis testing was conducted using HLM 6.0 as the analysis software.[*] The relative influence of LM on operating performance (OP) represents LM effectiveness. For each country, a facility's effectiveness was estimated and modeled to depend on the GLOBE cultural dimensions. HLM allows the statistical coefficients representing LM effectiveness to vary to model a culture's moderation and accounts for the fact that facilities are embedded within a country context (i.e., nonindependent facilities). In addition, HLM accounts for large variations in sample sizes among countries to avoid biased results[†], an important aspect of cultural studies.[‡] This is why HLM is a more robust method for this study than a multiple-regression method that must rely on interaction terms to show moderation.[§] Data were grouped into two sets: The

[*] S.W. Raudenbush, A.S. Bryk, Y.F. Cheong, R. Congdon, and M. du Toit, *HLM 6: Hierarchical Linear and Nonlinear Modeling*, Scientific Software International, Lincolnwood, IL (2004).

[†] S.W. Raudenbush and A.S. Bryk, *Hierarchical Linear Models*, 2nd ed., Sage, Thousand Oaks, CA (2002).

[‡] Kull and Wacker, Quality management effectiveness (2010); S. Wong, M.H. Bond, and P.M.R. Mosquera, The influence of cultural value orientations on self-reported emotional expression across cultures, *Journal of Cross-Cultural Psychology* 39 (2), 224–229 (2008).

[§] D.A. Hofmann, An overview of the logic and rationale of hierarchical linear models, *Journal of Management* 23 (6), 723–744 (1997).

first, level 1 set contains GMRG responses for facility *i* within country *j* for both LM and OP, and the second, level 2 set contains GLOBE culture value scores for country *j*. Although only six cultural dimensions were hypothesized to have an effect, all nine GLOBE dimensions are included for completeness.

Two control variables were also added, one to each level. At the facility level, organizational size (SIZE) was measured by the number of employees and controls for the greater difficulty in coordinating large companies.* At the country level, per capita gross domestic product (GDP) controlled for a country's degree of development, which reflects market competitiveness, technological readiness, and strength of standards. While a particular facility-level culture was likely present, and although it was not captured in the GMRG data, our method partitions out unique facility-level influences and thus captured the variance of interest in this study. This approach accounts for unique influences at the facility level.

The random coefficients with the level 1 covariate model are as follows:

Level 1: LM Effect†

$$OP_{ij} = \beta_{0j} + \beta_1 \left(SIZE_{ij} \right) + \beta_{2j} \left(LM_{ij} \right) + r_{ij} \tag{4.1}$$

Level 2: Country Effects

$$\beta_{0j} = \gamma_{00} + \gamma_{01} \left(GDP_j \right) + u_{0j} \tag{4.2}$$

$$\beta_1 = \gamma_{10} + u_1 \tag{4.3}$$

$$\beta_{2j} = \gamma_{20} + \gamma_{21} \left(IC_j \right) + \gamma_{22} \left(UA_j \right) + \gamma_{23} \left(AS_j \right) + \gamma_{24} \left(FO_j \right) + \gamma_{25} \left(HO_j \right) + \gamma_{26} \left(PD_j \right) + \gamma_{27} \left(GC_j \right) + \gamma_{28} \left(GE_j \right) + \gamma_{29} \left(PO_j \right) + u_{2j} \tag{4.4}$$

* T.J. Douglas and L.D. Fredendall, Evaluating the Deming management model of total quality in services, *Decision Sciences* 35, 393–422 (2004).

† Note the "empty" model is a random intercept-only model with $Y_{ij} = \beta_{0j} + r_{ij}$ and $\beta_{0j} = \gamma_{00} + u_{0j}$.

In Equation 4.1, the variance of relative facility OP is explained by a random intercept varying by country β_{0j}, a nonvarying facility size effect β_1, the relative influence of LM varying by country β_{2j}, and error r_{ij}. A country's average OP in a facility with typical levels of LM is represented by the parameter β_{0j}, which is explained in Equation 4.2 by an intercept parameter, GDP and error. No cultural dimensions are included in Equations 4.2 or 4.3 in order to be consistent and focus on the culture-as-moderator perspective.[*] However, culture is included in Equation 4.4, where the influence of LM β_{2j} is explained by a culture-neutral impact of LM γ_{20}, cultural influences represented by γ_{21} through γ_{29}, and error u_{2j}.

To test hypotheses 1 through 7, our HLM used a generalized least squares procedure to estimate gamma parameters and weighed the level 2 regression in favor of countries with higher level 1 estimation precision.[†] In this way, parameters in the level 1 and level 2 equations were estimated for each country and combined through an empirical Bayesian procedure to optimally weigh the level 1 estimates and the level 2 predicted values for these same estimates. Thus, the HLM provided the value and significance levels of each parameter to test hypotheses.

Cross-country studies commonly have a low sample of countries,[‡] causing the level 2 degrees of freedom to be low and thus making the detection of level 2 effects difficult.[§] A backward model-fitting approach handles this difficulty.[¶] Therefore, all nine cultural dimensions and GDP were initially included, but then the model followed a systematic removal of the most statistically insignificant variables until only a final set of significant coefficients remained.

SUMMARY

The method used to establish the relationship between LM effectiveness, operational results, and cultural influence must be based on reliable and

[*] Snijders and Bosker, *Multilevel Analysis* (1999).

[†] Hofmann, An overview (1997).

[‡] B.B. Flynn and B. Saladin, Relevance of Baldrige constructs in an international context: A study of national culture, *Journal of Operations Management* 24, 583–603 (2006).

[§] D.A. Ralston, D.H. Holt, R.H. Terpstra, and Y. KaiCheng, The impact of national culture and economic ideology on managerial work values: A study of the United States, Russia, Japan, and China, *Journal of International Business Studies* 28, 177–207 (1997).

[¶] Because OP is relative primarily to domestic competitors, LM is country centered, and GDP and culture dimensions are grand centered for interpretation consistency (Snijders and Bosker, 1999).

generally accepted research methods. Much of the mythology of LM and Lean systems is based on third-hand anecdotal evidence or unvalidated observational information. The GMRG and GLOBE study information provides an extensive data set on which to base the hypotheses of LM effectiveness and the cultural influences that have an impact on it. Through this approach, and by employing a data validation technique, there is a high degree of confidence in the results and the managerial actions stemming from those results.

While the data analysis model may not be immediately meaningful for the LM practitioner, it is important that these frontline Lean resources recognize their role in validating the potential cultural influences on Lean implementation. By understanding the methods by which LM effectiveness and practices might be compromised in various cultural settings, the practitioner can be assured of a greater level of performance success. Also, simply stated, this methodological approach is consistent with the statistical foundation of decision making and process adjustment seen in LM applications. The form fits the function.

QUESTIONS TO CONSIDER

1. How do you understand the employee cultural influence as a basis for statistical analysis? Do you see it as a controllable influence, or is it perceived as “noise”?
2. Can your company quantify if LM is more effective in one facility versus another? How could you do so, and can you “correlate” effectiveness with possible reasons why?
3. Does your organization know how to use cultural data to help ensure the successful implementation of LM?
4. Are the statistical methods that are used by your company “normalized” for cultural bias? That is, do you know how to account for the cultural influence on performance?
5. How is your company comparing the amount a facility invested in LM with the competitiveness of that facility? That is, are you able to measure the effectiveness of LM?
6. Can “second-order” data reduction be used within your LM environment to “distill” data performance factors into accurate depicters of broader LM effectiveness?

7. What lessons for the GLOBE study are culture can be applied in your company, especially across national and geographic regions?
8. Is your facility's culture similar to the culture of the country where it is located? How does the country culture influence the mentality of employees, and does that matter to your operation?

5

Assessing the Results

DATA ANALYSIS SUMMARY

As with Lean manufacturing (LM) process statistics themselves, it is necessary to analyze and explain the survey result data. And, like process statistical process control (SPC), there are no convenient shortcuts or assumptions in interpreting the results. For those not inclined to wade through the data, Table 5.1 is a brief summary of data results as they relate to the LM effectiveness hypotheses.

WHAT DID THE DATA SHOW?

To understand the results, it is necessary to understand the method to interpret the results, just as it was important to understand the survey as explained in the last chapter. Table 5.2 shows the results of the hierarchical linear model (HLM) analysis. To provide an estimate of base variance in operational performance (OP), it is necessary to include an "empty" model that excludes any explanatory variables. The base operational variance in this empty model includes variance for both within-country σ^2 and between-country τ_0^2. As variables are added to the analysis model of the respective OP characteristics, the percentage variance explained R^2 is adjusted to reflect the percentage reductions in error variance.

Table 5.2 first compares the empty model to model 0; model 0 included fixed control variables of organizational size (SIZE) and per capita gross

TABLE 5.1

Summary of Data Relevant to LM Effectiveness Hypotheses

Hypothesis	Anticipated Result	Actual Result
H1: LM + to OP	LM will have positive effects on OP.	**Confirmed**: Hypothesized results verified by the GLOBE data study
H2: IC + to OP	Institutional collectivism will have a positive effect on OP.	**Refuted:** Hypothesized results refuted by GLOBE data; results indicate no appreciable effect
H3: UA + to OP	UA positively affects OP.	**Confirmed:** Operational performance positively affected by high UA
H4: AS (−) to OP	OP is negatively affected by high AS.	**Confirmed:** LM effectiveness negatively moderated by high AS
H5: FO (−) to OP	Future orientation within a culture has a negative effect on LM effectiveness.	**Confirmed:** Study data demonstrated significant negative influence of FO on OP
H6: HO (−) to OP	HO negatively affects operational performance.	**Refuted:** No appreciable negative moderation by HO on operational performance
H7: PD (−) to OP	PD negatively moderates LM effectiveness.	**Refuted:** Analysis of GLOBE data showed no appreciable effect on LM effectiveness
No effect 1: GC	GC has no appreciable effect on OP.	**Confirmed:** No appreciable effect noted in GP
No effect 2: GE	GE does not affect OP.	**Confirmed:** GLOBE survey analysis showed no notable effect of GE on OP
No effect 3: PO	PO has no effect on OP performance.	**Refuted:** Data showed that cultural PO actually had a strong positive effect on operational performance

domestic product (GDP). Controlling for SIZE and GDP was statistically significant in model 0, which showed that OP was relatively higher in less-developed countries (lower GDP) and in larger organization. Model 0 also showed a relatively high between-country OP variance.* The data showed that model 0 reduced facility-level variance σ^2 influence by 7.6% and country-level influence τ^2 by 37.7%. These variances indicated that SIZE explained little of the facility-level OP variation, while GDP explained a substantial portion of the country-level OP variation.

* The reliability estimate for the randomly varying intercept parameter was 0.858, indicating the high between-country OP variance.

TABLE 5.2

HLM Results

	Dependent Variable: Operating Performance				
Parameters	**Empty Model Estimate (SE) [*t* value]**	**Model 0 Estimate (SE) [*t* value]**	**Model 1a Estimate (SE) [*t* value]**	**Model 1b[a] Estimate (SE) [*t* value]**	**Model 2 Estimate (SE) [*t* value]**
Grand intercept					
$\gamma_{00}{}^{b}$	5.019 (0.075) [66.567]***	5.011 (0.060) [83.347]***	5.019 (0.059) [84.132]***	5.021 (0.059) [84.004]***	5.021 (0.059) [84.267] ***
Control variables					
γ_{01} (*GDP*)		−0.001[b] (0.001) [−3.891]***	−0.001[b] (0.001) [−3.937]***	−0.001[b] (0.001) [−3.634]***	−0.001[b] (0.001) [−3.916]***
γ_{10} (*SIZE*)		0.001[b] (0.001) [1.808]*	0.001[b] (0.001) [0.272]	0.001[b] (0.001) [−0.238]	0.001[b] (0.001) [−0.408]
Hypothesized effects					
γ_{20} *LM* (H1 +)[c]			0.221 (0.018) [12.007]***	0.224 (0.028) [7.860]***	0.237 (0.019) [12.137] ***
γ_{21} *IC* (H2 +)					n.s.
γ_{22} *UA* (H3 +)					0.064 (0.036) [1.778] *
γ_{23} *AS* (H4 −)					−0.083 (0.038) [−2.193]**
γ_{24} *FO* (H5 −)					−0.120 (0.064) [−1.875]*
γ_{25} *HO* (H6 −)					n.s.
γ_{26} *PD* (H7 −)					n.s.
Other cultural effects					
γ_{27} *GC*					n.s.
γ_{28} *GE*					n.s.
γ_{29} *PO*					−0.337 (0.082) [−4.113]***

continued

TABLE 5.2 (continued)

HLM Results

	Dependent Variable: Operating Performance				
Parameters	**Empty Model Estimate (SE) [*t* value]**	**Model 0 Estimate (SE) [*t* value]**	**Model 1a Estimate (SE) [*t* value]**	**Model 1b[a] Estimate (SE) [*t* value]**	**Model 2 Estimate (SE) [*t* value]**
Deviance (*D*)	3350.5	3379.2	3246.1	3237.1	3227.9
σ^2	0.562	0.561	0.510	0.501	0.499
τ	0.124	0.073	0.073	0.074	0.073
τ				0.009	<0.001
Reliability					
β_{0j}	0.908	0.858	0.868	0.871	0.871
β_{2j}				0.507	0.070
% MSE reduction Y_{ij} (R_1^2)	n/a	7.6%	14.9%	16.1%	16.5%
% MSE reduction β_{0j} ($R_{2,0}^2$)	n/a	37.7%	37.8%	37.6%	38.0%
% MSE reduction β_{2j} ($R_{2,2}^2$)					56.1%

Note: n/a, not applicable; n.s., not significant; SE, standard error.

[a] Based on model 1a with random LM effectiveness u_{1j} added for comparison with model 2.

[b] Values below 0.001 are shown as 0.001.

[c] Shows LM effectiveness not affected by culture.

[*] $p < 0.10$, [**] $p < 0.05$, [***] $p < 0.01$.

OVERALL EFFECTIVENESS OF LEAN MANUFACTURING

The effectiveness of LM was shown in models 1a and 1b. These two models included the same standardized control variables as model 0: country per capita GDP and organizational size. In addition, the models examined the hypothesized effects of LM effectiveness on OP. In model 1b, the effect of cross-country influence was shown with respect to LM effectiveness. The purpose of this one-dimensional variance analysis was to provide a basis for comparison to model 2, addressed further in the chapter (model 2 was a fuller evaluation of cultural dimension influences).

When the hypothesized effect of LM effectiveness was included as part of the GLOBE (Global Leadership and Organizational Behavior

Effectiveness) data analysis, the results of both models 1a and 1b indicated that SIZE was no longer statistically significant (as in model 0). The change in this statistical significance indicated a potential positive relationship between LM and organizational size (SIZE). In layperson's terms, this means that when LM was excluded from the analysis, larger-size organizations showed greater OP. When the influence of LM was added, the size effect diminished, meaning that the higher OP was determined by LM, not by SIZE. The positive relationship between LM and SIZE indicated that larger organizations tended to include LM systems. Also, models 1a and 1b supported the fundamental assumption and hypothesis that LM positively influences operation performance* ($\gamma_{20}^{1a} = 0.221$, $p < 0.001$ and $\gamma_{20}^{1b} = 0.224$, $p < 0.001$). While this "proof" was of no great revelation to those who have integrated LM and Lean systems into their operational disciplines, the data-based analysis did provide the foundational evidence that LM is not merely a trendy manufacturing fad but a fact-based efficiency system. The results suggest that LM can be effective in any country regardless of culture. Yet further analysis will show where cultural influences make LM more or less effective. Less effective does not mean "ineffective"; it means less effective than other locations.

These basic three-factor models (GDP, SIZE, and LM effectiveness) supported hypothesis 1 (H1) that LM positively influences OP. This was supported further in the subsequent models, which began to incorporate more of the hypothesized elements. The reliability in detecting a country's true OP increased in these subsequent models. This increase in reliability indicated that as LM effectiveness was accounted for, more between-country differences in OP existed. For the LM manager, this means that any potential ineffectiveness in LM deployment was more likely attributed to individual country cultural elements than due to failures in the Lean methodologies themselves.

Finally, this initial set of data shown in models 0–1b indicated a substantial increase in explained percentage variance R^2 when the LM factors were added to the analysis model. The overall data error as calculated by the mean square error (MSE) was reduced by nearly 15% in model 1a and by over 16% in model 1b. This change in the level 1 R^2 values showed that facility-level OP was significantly explained by the effects of LM disciplines.

* The regression coefficients were unstandardized following typical HLM output.

EFFECTS OF COUNTRY CULTURAL DIMENSIONS

Model 2 expanded on the 0–1b models by adding the cultural value dimensions. This model made use of the "backward model-fitting" procedure, a process by which dimensions that were statistically insignificant were removed from the data analysis and noted with the annotation "n.s." Therefore, as indicated in Table 5.2, institutional collectivism (IC) was statistically insignificant in the data analysis and therefore undermined hypothesis 2 (H2) that LM effectiveness will increase in the global cultures that value social mechanisms for rewarding collectivist behavior. Again in layperson's terms, this means that LM effectiveness has no appreciable difference in highly controlled or socially structured cultures. A LM manager cannot anticipate higher rates of improvement in OP due to LM in those cultures that tend to "do as they are told" and typically exhibit less resistance to modified operational systems.

However, the effects on OP were significant ($\gamma_{22} = 0.064$, $p < 0.10$) with respect to the influence of uncertainty avoidance (UA) within a given culture. Hypothesis 3 (H3) was validated by this finding. H3 stated that UA positively affects LM effectiveness, so what we see is that in those cultures and countries that value predictability, the application of LM systems improve OP. These cultures more willingly adapt to the application of the LM "stepwise" approach to quality, process, and system improvement and embrace the frequent feedback mechanisms within LM. The frequent feedback mechanisms provide the methods by which increasing certainty can be obtained in the manufacturing environment and as a result fit within the cultural views of such predisposed countries. Lean managers should expect a warm embrace of the LM methodologies that provide these feedback mechanisms.

Our original hypotheses included the expectation that four cultural values would negatively influence LM effectiveness. Those four cultural values are assertiveness (AS), future orientation (FO), humane orientation (HO), and power distance (PD). Each of these factors was included in the model 2 analysis. The model 2 analysis showed that, as postulated, AS did have a moderate negative effect on LM effectiveness ($\gamma_{23} = -0.083$, $p < 0.05$). Even with moderate effect, the negative influence of AS supported hypothesis 4 (H4): The degree of assertiveness in the culture of the country of a manufacturing facility negatively moderates LM effectiveness.

This finding suggests that those cultures that value individual responsibility and competition are resistant to or unable to integrate those work structures, processes, or systems that are based in collaboration and cooperation. Workers in these cultures have difficulty seeing process errors, production failures, or quality problems as systemic to the manufacturing processes and systems (vs. individual failure). These workers also struggle to develop cooperative relationships with other workers and supervisors. This may often be seen in an "us-versus-them" attitude and be prevalent in contentious or hostile unionized environments.

The second hypothesized negative effect on LM regarding the influence on operational effectiveness was the dimension of FO. Hypothesis 5 (H5) proposed that cultures with greater FO exhibit reductions in LM effectiveness. Model 2 data analysis showed that there was indeed a significant negative affect on LM effectiveness ($\gamma_{24} = -0.120$, $p < 0.10$). In fact, the negative influence of FO was 50% greater than that of AS. From a managerial standpoint, this finding indicates that those cultures that stress long-term planning in pursuit of a predictable ideal state beyond the present environment will have significant difficulty in maintaining the appropriate state of urgency with respect to immediate process performance. This is in counterpoint to low-FO cultures, which believe that the future is unpredictable; therefore, these cultures are more sensitive to minor changes in the immediate environment. What this means is that the lack of urgency in high-FO cultures coupled with the desire (and practice) of long-range planning will degrade many of the key features and practices of LM and Lean systems.

The final two cultural values, HO and PD, were also hypothesized to have a negative effect on LM effectiveness. These hypotheses were not validated by the model 2 findings. As indicated in the data, both of these factors were insignificant with respect to explaining any influence or cultural impact of LM effectiveness on OP. Hypothesis 6 (H6), that the degree of HO in the culture of the country of a manufacturing facility negatively affects LM effectiveness, was not supported. Hypothesis 7 (H7), that the degree of PD in the culture of the country of a manufacturing facility negatively affects LM effectiveness, was not supported. These findings indicated that the level of a culture's concern with worker rights or those cultures that prioritize formal hierarchies were equally supportive (or amenable) to LM structures and initiatives as those countries that did not possess such concerns.

"NO-EFFECT" CULTURAL DIMENSIONS

There were also three cultural dimensions that, based on the published literature regarding LM, were hypothesized to have no effect on LM effectiveness or improving OP. Those three dimensions were in-group collectivism (GC), by which workers seek the security, loyalty, and pride in work and other social groups; gender egalitarianism (GE), which addresses the degree of gender equality within a culture and work environment; and performance orientation (PO), a cultural dimension that encourages its members (or workers) to excel against the performance of peers.

As anticipated, the cultural dimensions of GC and GE effects on LM were not significant in the study data. This may be an important social factor as well as a cultural influence as the findings suggested that regardless of a country's perspective and action on issues of gender equality or group loyalty, LM will be equally effective. As we indicated when we described these cultural elements in Chapter 3, GE is described in the professional and academic literature not simply as neutral or having no effect, but actually having counteracting effects. To the LM manager, this distinction is important because instead of being an element that can be largely ignored, the manager may need to ensure that any individual negative elements are mitigated or offset with those having positive influence. Despite these factors being insignificant with respect to LM effectiveness, it does not mean that these elements can go unmanaged.

Perhaps the most surprising result indicated in the GLOBE study data was the effect of PO. The model 2 analysis showed a surprisingly strong correlation between high PO and negative effect to LM ($\gamma_{29} = -0.337$, $p < 0.01$). This finding implies that those cultures that indicated a strong self-initiative and productivity focus may actually be subverting the principles of LM rather than embracing them. Again, the professional literature suggested a counteracting effect of the elements within this cultural dimension. Our hypothesis was derived from the literature and practitioner survey data.* The Lean manager, and in fact the company executives, should be greatly concerned with the over-emphasis on individual performance objectives and the fostering of a competitive performance environment as this may actually have unintended and undesirable

* Following the initial study, additional academic journal research indicated a stronger negative influence on LM effectiveness. In our subsequent study (under review), we reexamined our hypotheses and theory to come to a formal hypothesis for why PO should be negative.

consequences. Instead of promoting the desired breakthrough results that the company seeks, PO creates a drag on the effectiveness of LM techniques and systems. This negative effect can be especially problematic in a country like the United States, which prides itself on its results orientation and whose manufacturing companies have been perceived as expertly adopting LM. Successfully implementing LM in such a country culture will have to overcome a twofold ego barrier. First, a culture that suggests that performance orientation among manufacturing employees is the only (or best) method for success must be overcome.

The second ego barrier to overcome is a culture that believes its "head knowledge" of LM is successfully translated into an effective "body knowledge" of implementation. Head knowledge describes the intellectual internalization of LM and Lean systems within a company or organizational site; body knowledge describes the adaptation and mastery of LM in actual practice. Many cultures and individual organizations believe that the intellectual understanding and academic mastery of LM vision, structure, tools, and approaches successfully translates to operational deployment—without sufficient history or evidence of success. Such cultures often do not want to believe that it may not actually be as good as it believes and certainly must be better than others.

STUDY SUMMARY

The various control and hypothesized variables included in the different models increasingly clarified the cultural effects on LM effectiveness and OP. Model 2 provided the clearest examples regarding why LM may or may not work as well within a given country or culture. This can be seen in the walk from model 1b to model 2. Where model 1 included the control variables of GDP and company SIZE, along with LM, model 2 included those LM values experts believe affect the performance of LM. The MSE of model 2 showed a 56.1% reduction over model 1b. This means that the factors that describe the various aspects of positive or negative LM performance became more precisely defined when the hypothesized effects were added.

In this new model, there existed only a 0.4% improvement in predicting facility-level and company-level operating performance. So, while the control variables and LM factor (model 1b) present within a culture predicted nearly the same level of OP as those with the extended cultural elements (model 2), the more complete analysis reflected a significant decrease in the deviance of the study values. Because of this, it becomes possible to begin to predict the facility OP due to the implementation of LM. Local, regional, and country-level managers may be better attuned to the culturally based challenges of implementing LM, and therefore the benefit of understanding a country's cultural values is important when implementing LM.

QUESTIONS TO CONSIDER

1. Does your company size play a role in the ability to implement LM approaches?
2. Are there artificial barriers, such as artificially suppressing the organizational size or facility/site revenue (SIZE and GDP factors), put in place by your company to inhibit you from being successful in LM deployment?
3. Have you or your organization realized the positive effect to your LM system from UA? That is, do your LM efforts include the messaging of predictability of results?
4. How have you managed to counteract the negative influences on LM for AS and FO?
5. Does your organization possess the ability to recognize and encourage performance results without an overemphasis on individual competition? Can your company distinguish between systems-driven performance outcomes such as LM versus individualistic outcomes and know how to manage to the appropriate expectations?
6. When deploying LM to a new manufacturing location or country, are you now better able to determine the barriers and enablers that affect success? How similar is your plant's culture to your country's culture?
7. What message do you get when you read these results? How would you sum up these results to your boss or colleagues?

6

The Implications of Lean Culture

INTRODUCTION

As Lean manufacturing (LM) gradually becomes the gold standard of operational practice, practitioners and researchers increasingly attempt to describe how and why LM works and what factors might limit improvements in operational performance. Increasing globalization and continually expanding manufacturing footprints have allowed management teams to observe LM operational results from multiple country and region perspectives. This multinational view has provided greater insight regarding the enablers and barriers that drive LM and particularly employees in the various LM environments.

Capitalizing on the ever-expanding LM global deployment, many authors have argued that, based on the observed and reported performance from within global Lean operations, the success of LM depends on deploying LM within a culture that is *congruent with the values underlying LM*. Many of the authors, however, based these proposals on supposition, anecdote, and limited observation, offering performance theories based on what seems obvious and reasonable.

What we believe has been necessary is a critical examination of the relationship between cultural values, LM values, and LM effectiveness. The connection of and between these three elements provides actionable effects from which LM managers are able to proactively tailor local LM implementation strategies and efforts. The eight LM values derived from the existing literature (as we described in Chapter 2) form the foundation of understanding *what* LM is supposed to be. The six hypotheses[*] presented in Chapter 3 form the basis of *how* cultural biases affect the effectiveness of LM. Using the global manufacturing data, four of those hypotheses held true, while two showed little or no effect. This data approach not only provides the next level of validation from the previous authors' validation of LM and cultural moderation but also provides a clearer basis of understanding as to why some cultures or regions may be incompatible to a "generic" adoption of LM and other Lean systems. Essentially, we have found that some cultures *will* adopt Lean (or attempt to adopt Lean), but it will not be effective. The implication is manufacturing sites in these countries/cultures have implemented Lean practices incorrectly, that they are using Lean practices incorrectly, or that there are other cultural factors deterring Lean from being beneficial. The theory pointing toward "using it wrong" is the strongest—but the other explanations are also possible. Most important, however, the findings of this approach become powerful tools with which managers are able to develop country contingencies for LM implementation. These tools enable the managers to more precisely define the *form* of LM and the *method* of LM implementation.

The eight LM values we have used are shown to be beneficial in predicting the cultural implications of LM implementation. This synthesizes previous descriptive accounts of the cultural underpinnings of LM using the eight value dimensions.[†] The pattern of results suggests certain LM values are more salient than others. The LM values that appear most influential relate to motivation, variance, vocation, and cooperation—each internal integration mechanism that drives culture.[‡] The congruence among this set of values and each cultural dimension appears to best predict the direction of the moderating effect. Just as all the LM practices are interrelated to make LM an integrated system,[§] these four LM values form a coherent value system underlying LM. That is, in the process of continuously removing waste to realize the "ideal" production system (LM4, LM5), buffers in time, capacity, and inventory are gradually removed, creating a high-tension system whose elements are highly interrelated (LM6). To solve systematic problems (LM3) to increase productivity (LM5), interunit

[*] In their working paper version 2 dated January 24, 2012, Drs. Kull, Yan, and Wacker expanded their original six theses to include a seventh hypothesis that addresses the hypothesized effects of performance orientation-negative mediation on LM effectiveness.

[†] Detert, Schroeder, and Mauriel, A framework for linking (2000).

[‡] E.H. Schein, *Organizational Culture and Leadership*, 3rd ed., Jossey-Bass, San Francisco (2004).

[§] Shah and Ward, Defining and developing measures (2007).

collaboration (LM6), distinct work duties (LM5), and stepwise improvements (LM4) are all necessary.

THEORETICAL IMPLICATIONS

Of the four hypotheses validated by the GMRG (Global Manufacturing Research Group) manufacturing study data, one was found to have a positive effect on LM effectiveness; three were found to have a negative effect. The results showed that cultures high in uncertainty avoidance (UA) *enhanced* LM effectiveness; those with high future orientation (FO), high assertiveness (AS), and high performance orientation (PO) *diminished* LM effectiveness.*

These findings suggest that LM effectiveness will be greater or enhanced in those countries whose cultures value predictability. In these cultures, workers possess an urgent need or desire to learn to correct current problems in a cooperative, nonconfrontational manner. These problems are corrected by addressing systemwide causes. Manufacturing employees in these countries exhibit cognitive consonance (peace of mind) in environments that are more predictable, and as such they are more likely to accept LM methodologies. Their acceptance of LM may be based on the "uncertainty-reducing" nature of continuous monitoring, detailed routines (via standardized work), frequent feedback, and stepwise improvement common to all LM practices. In these countries, the intersection of the cultural values and LM values results in a more "voluntary" acceptance of LM.† As a result, there is less corruption to LM implementation and methods in all aspects of operational deployment.

In contrast to the enhancing effect of UA, FO diminishes LM effectiveness. The incongruence between FO cultures and LM values lies mainly in the "ignorance" (or ambivalence) such cultures exhibit in recognizing immediate signals and targets. Workers in these countries typically do not possess a sense of urgency in achieving short-term "target conditions" or

* In all cultures/countries, LM has a positive effect on OP as shown in H1. As a result, characteristics that enhance or diminish LM effectiveness should be understood to mean that the amount of improvement in OP is either enhanced or diminished, but in no case is it eliminated.

† An alternative way of looking at this adoption question is to consider the culture "fully embracing" LM in a manner that is nonsuperficial or noncorrupting. *Voluntary* is used to describe a noncoerced adoption of LM, where the employees' values are more aligned with LM values, thus resulting in greater benefit to operational performance.

in responding to frequent system feedback. Because of this, the conditions do not exist where LM systems and system processes can be effectively monitored or where continuous improvements could be made.

Among the fundamental Lean system assumptions is that LM practices enable processes that provide for systemwide improvements. In assertive cultures, an overemphasis on defending individual opinions in work environments where employees compete for personal benefits makes the development of systemwide improvements far more difficult. Even the unexpected influence of PO cultures belies the effect of assertive cultural influence has on LM effectiveness. While an a priori understanding of cultures with high PO would/could imply that individual performance extends to enterprise-wide improvement, the data analysis showed a negative effect on LM effectiveness. This finding seems to imply that the drive for "self-improvement" is an individual trait and therefore does not transfer to the organizational level. So, while individual performance improvement may be a cultural motivator, this "urgency" does not apply to group or systemwide enhancement. While often missed by operational managers, the importance of distinguishing between individual-, facility-, and and country-level characteristics is critical in understanding motivations for improvement.*

ORGANIC WORLD COMPLEXITY

Perhaps the most difficult task in understanding and applying the cultural influences on LM effectiveness and operational performance is that the influences and applications exist in an organic and ever-changing environment. That is, the management of a production system can be viewed as a difficult task in which an ever-changing internal and external environment creates unpredictable forces that need constant monitoring, experimentation, and learning. Predictability is always threatened, and full understanding never exists. In such an ever-changing global backdrop, there are very few countries, in fact, that exhibit the "ideal"

* E. Miron, M. Erez, and E. Naveh, Do personal characteristics and cultural values that promote innovation, quality, and efficiency compete or complement each other? *Journal of Organizational Behavior* 25, 175–199 (2004).

combination of all the cultural contingencies that have a positive effect on LM effectiveness.

Companies with multinational manufacturing presence then face a particularly challenging paradigm in which expectations of Lean implementation across cultures are more difficult than expected. For example, the cultural biases in China reflect high UA and low FO, both of which have a favorable effect on LM; yet, Chinese culture also exhibits high AS, which negatively moderates LM effectiveness.* Germany also indicates a complex combination of cultural traits, yet in a substantially different combination from China. German culture reflects LM-enhancing traits of low AS and FO, with diminishing effect of low UA. In Chapter 7, we investigate broader cultural combinations to better understand the effect and accommodations that need to be considered to effectively deploy LM.

Assuming certain cultural biases have a better likelihood in promoting LM effectiveness is a common mistake among Lean practitioners. In particular, many make unsubstantiated assumption of Japanese or Japanese-like cultures primarily driven by the observed success to Toyota. As the benchmark of LM systems, the Toyota Production System (TPS) is typically equated with Japanese culture. However, while TPS and the Toyota culture possess similarity to LM values, Toyota's culture does not necessarily represent, or is not entirely representative of, Japanese culture. For example, Japan highly values AS, a cultural trait that has been shown to diminish LM effectiveness. Toyota has intentionally located its headquarters away from mainstream culture in Japan, to the relatively remote rural area in Aichi Prefecture.† The purposeful location of its headquarters away from the typical influences of Japanese culture may be to counteract the influences that negatively affect its operational performance. Simply said, because the ideal mix of cultural contributors of Japanese culture varies from those indicated in our research model, Japan or cultures similar to that of Japan should *not* be assumed to be optimal or more effective.

In researching their study on quality management effectiveness, Kull and Wacker only found two cultural dimensions influential on performance.‡ The research presented in this book now shows four cultural dimensions having notable influence on operation performance. What this implies is that although it may be considered an extension of quality management,

* Gupta and Hanges, Regional and climate clustering (2004).
† Rother, *Toyota Kata* (2009).
‡ Kull and Wacker, Quality management effectiveness (2010).

LM has a broader reach on operation performance and is more sensitive to culture and cultural influences. The sensitivity to cultural differences by LM may not be surprising based on the relative "newness" of LM in may operational settings, as well as its prescriptions for how people learn and act. These two traits are highly influenced by culture.[*] The well-known Lean process expert Mike Rother has spent many years researching the TPS. He believes the improvement *kata*, a well-rehearsed routine-based method of improvement, is a key to the successful implementation of LM practices.[†] As a new method of thinking and acting within the manufacturing environment, the improvement kata is a method with characteristics sensitive to culture. There has been much attention given recently to these Lean practices (especially newer perspectives) interacting with cultural influences.[‡]

The final theoretical consideration is to be cognizant of the fact that LM is not unique in being subject to or experiencing the effects of cultural differences. Because other functional and operational departments and disciplines are equally affected by various country cultures, they have a notable effect on the form and success of a company's systemwide approach to LM and Lean systems.[§]

Cultural biases will also be a significant determinant in the behavior of facility leadership. Country culture will affect whether facility leaders will emulate LM disciplines and adopt practices such as "standard work for leaders."[¶] Operational and other functional leaders who practice a leader's standard work will accept and participate in activities such as routines of improvement and regular *gemba* walks, subject to the cultural influences we have been discussing.

Perhaps somewhat unfortunately, the majority of discussions and literature on LM focused almost exclusively on production operators. This focus misses the fact that country culture has a facility-wide effect on what is considered to be the "shared system" of assumption, value, and beliefs.[**] What we have found in this research, and what we have been discussing throughout this book, underscores this point. As we have shown,

---

[*] Schein, *Organizational Culture and Leadership* (2004).

[†] J. Liker and M. Rother, *Why Lean Programs Fail*, Lean Enterprise Institute, Cambridge, MA (2011).

[‡] See recent posts about Lean culture in the Lean Weekly Forum Digest. http://www.lean.org.

[§] Browning and Health, Reconceptualizing the effects (2009); Liker and Hoseus, Human resource development (2010).

[¶] Also called leader standard work. See various discussions at http://www.lean.org.

[**] Schein, *Organizational Culture and Leadership* (2004).

the influential nature of country culture and understanding the cultural processes that explain the variation in LM effectiveness (and ultimately operational performance) substantially benefit both organizational management (OM) research and Lean deployment. While the theoretical implications may be valuable to academics, it will be the managerial implications that we discuss next that will provide the greatest value to Lean practitioners.

MANAGERIAL IMPLICATIONS

There are a number of key managerial take-aways with respect to cultural influence and LM deployment. The first of these is that not only are there cultural influences on LM effectiveness and operational performance, but also the extent of "improvement" or enhancement of LM effectives is directly affected by the extent of LM investment. The GMRG data and analysis demonstrated the positive impact of LM on operational effectiveness and increasing relationship/correlation between the two. What this means is that the greater and more extensive the investment in LM techniques and practices, the greater the improvement in operational performance will be. In this case, investment does not simply mean the financial investment or capital committed to LM systems but also to the depth and extent of organizational and cultural investment, investments in the process, discipline, results, outcomes, and so on. Figure 6.1 shows the relationship between LM integration and quality performance.

The second important managerial take-away is that the decision criteria by which country and facility choices are made will have significant influence on the pace and magnitude of cultural influences. The previously mentioned study by Kull and Wacker* gave an illustration of how increasing the investment in a management initiative (in that case a quality management initiative) will have differing payback depending on the country's cultural makeup. As seen in Figure 6.1, the combinations of various cultural influences create varying impact and acceleration of operational performance metrics. In this case, the rate and magnitude of quality management was evaluated against the cultural influences of UA

* Kull and Wacker, Quality management effectiveness (2010)

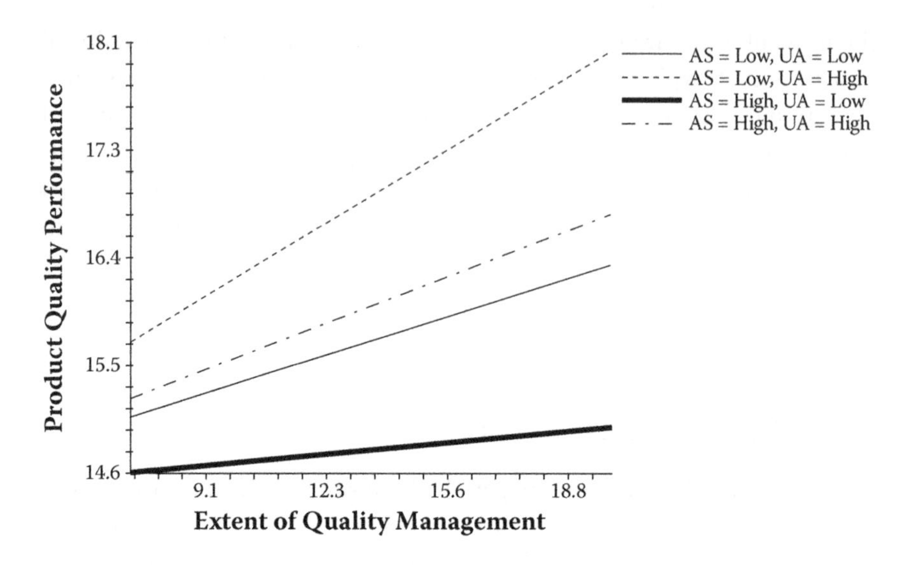

FIGURE 6.1
Combinations of cultural influences affect impact and acceleration of operational performance metrics.

and cultural AS. The improvement in quality performance is plotted against these factors.

Within the cultural four possible combinations of these two cultural dimensions (low AS/low UA; low AS/high UA; high AS/low UA; and high AS/high UA), each combination results in varying levels of increased product quality performance based on extent of quality management implementation. The lowest-effect combinations are those countries that have high AS and low UA. In these countries, the level of integration of quality management into the manufacturing organization does not result in appreciable improvement in achieved quality results. By contrast, those countries with low AS and high UA show significant improvement in quality performance as the extent of quality management integration increases. The slope of each line reflects the greater or lesser moderation that the cultural elements have on quality management's effect on operational performance.

Similarly, the research we have been presenting regarding LM showed the effect of certain cultural biases. As seen in Figure 6.2, the various cultural influences had an impact on LM effectiveness differently. Each bar represents the range of how LM improved performance given a cultural trait. The upper left chart shows how low-to-medium levels of UA generally had lower levels of LM effectiveness than high levels. The other three charts show the opposite, with higher levels of AS, FO, and PO displaying

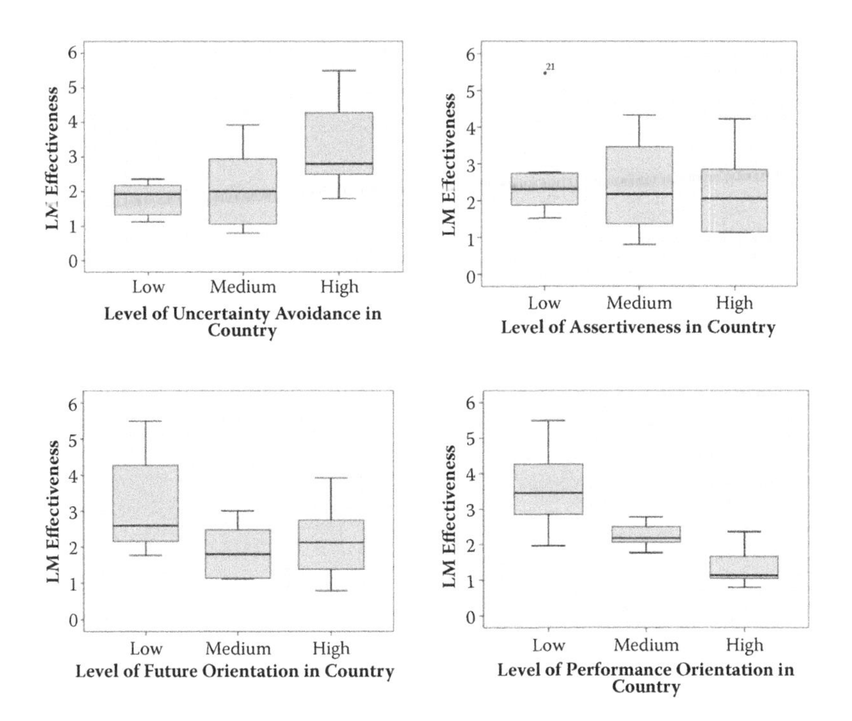

FIGURE 6.2

less LM effectiveness in general. Particularly noteworthy is that some cultural values were more narrowly grouped with the countries exhibiting a specific cultural value. For example, note that the effectiveness of LM varied less under certain conditions, such as in countries with low UA and higher PO. Another important point is that substantial variance existed within each high/medium/low grouping. This means that while the overall picture shows country culture dimensions to be influential, other influences like industry and facility leadership also exist. Another way to look at the in-country affect and distribution is as shown in Figure 6.3.*

These figures show each country's varying facility-level LM effectiveness, grouped by the country's cultural bias. Again, we see that substantial variation existed both within and across countries given a specific cultural bias. This is partly because of the other cultural influences at play, but also because of the unique characteristics and choices of each manufacturing facility. One interesting observation is that, for the most part,

* The range ordering of the data is not meant to imply an increasing trend among the data samples. The data were merely ordered in this manner for convenience.

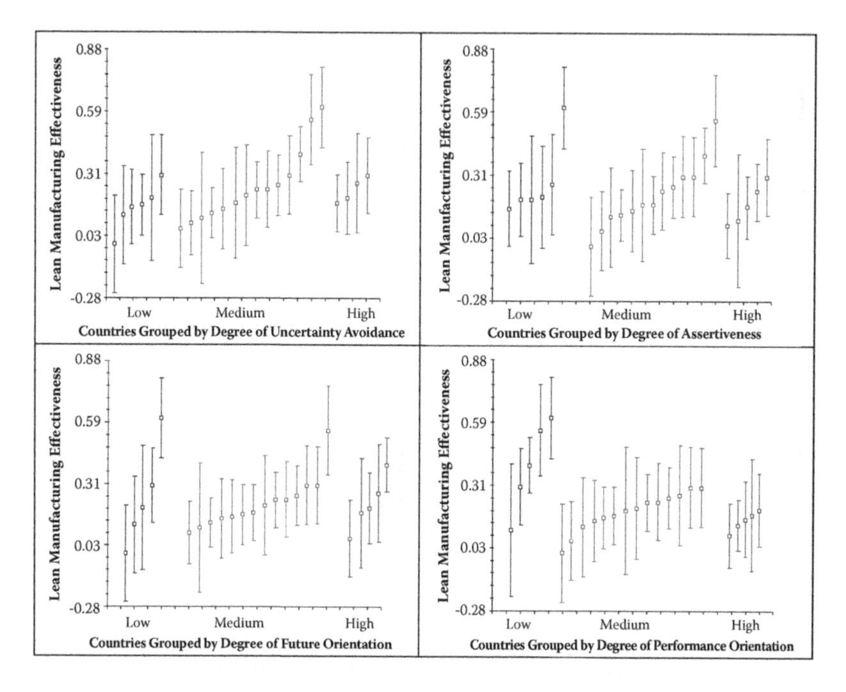

FIGURE 6.3

when a country highly values any particular dimension, the variability in LM effectiveness shrinks. This can be seen by comparing the width of the bar of "low" countries to the "high" countries. Cultural biases clearly seem more present when a dimension was highly valued than when it was not much valued. However, when looking at the general trend from low to medium, to high, the overall tendency and influence of that cultural dimension can be seen.

COUNTRIES OF CHOICE

Most decisions regarding expansion into foreign markets or expanding regions are based largely (or even solely) on financial considerations. Factors such as labor rates, property costs, market potential, taxation, and more generally drive facility decisions. Rarely do long-term productivity or efficiency differences among countries and regions come into play in facility location decision making.

With an eye toward operational performance, as companies make decisions regarding foreign facility development, managers should consider the benefits of national and organizational cultural influences that are favorable to LM. These nonfinancial performance considerations are particularly critical to those organizations that have already made substantial commitments and investments into LM and Lean systems or for those companies that have identified LM methods as critical go-forward strategies.

Among the considerations necessary to achieve the highest levels of LM effectiveness is understanding the crucial role of the internal and external supply chain and ensuring they follow LM practices. The certainty of adherence to LM practices and disciplines varies between countries and cultures. The countries that exhibit cultural behavior, values, and attitudes that are congruent with LM values provide a higher assurance that LM will be used as expected. The predictability of employee behavior within LM operations based on country and cultural values provide managers with the ability to choose global manufacturing locations and organizations that will be more adept at implementing and utilizing LM systems.*

The typical financial criteria are only one set of considerations that may direct a company to a particular country or region as it establishes new manufacturing sites. Other strategic or supply considerations may also drive location choices that are inconsistent with LM values. In these cases, managers may be able to dampen the negative effects on LM by being aware of the likely effects and planning mitigation strategies accordingly. For example, companies such as Lincoln Electric regularly adapt their managerial practices to better align to individual country differences. Lincoln responds to these country "contingencies" by implementing different elements of their management systems that best fit the individual or particular cultural reality.†

Examining this approach from an LM perspective, for example, using LM practices in high-FO cultures may mean an increase in the quantity of buffer inventories when using andon methods. Managers in these cultures

* While not explored in detail as part of this study, managers responsible for foreign facility decision making should also be aware of the migration effects of globalized labor forces. It may be equally important to understand the culture of origin of employees as well as the country of employment. Global worker migration may have other mitigating effects on LM effectiveness, potentially lowering the effect of the site location.

† J.I. Siegel and B.Z. Larson, Labor market institutions and global strategic adaptation: Evidence from Lincoln Electric, *Management Science* 55 (9), 1527–1546, (2009).

may need to factor in these increases as high FO also reflects a lower sense of urgency regarding immediate issues and problem solving. Similar to the way in which technical restraints may result in batch processing or buffer inventories, social constraints may necessitate similar buffer placement. Likewise, supplier accommodations can be made to lessen cultural influence. Local and international purchasing managers can prepare for differences in cultural realities, knowing suppliers in countries incongruent with LM values may deserve more attention.

HOW VERSUS WHAT TO IMPLEMENT

Most LM effectiveness mitigation or adaptation efforts such as the Lincoln Electric example focus on *what* LM elements or systems to deploy within a country or cultural region. However, equally important is the necessity to consider not only what to implement but also *how* to implement various LM disciplines. By knowing how as well as what to implement will give managers another competitive strength in achieving maximum LM effectiveness and operational performance.

An illustration of the "how" approach again looks at a culture with high FO. The "what" approach added buffer inventory to mitigate the effects of high FO and lack of urgency in problem solving. A how approach to mitigate the urgency issue would be to demonstrate to employees how reliance on inventory as a response to day-to-day issues ultimately creates future problems. Similarly, cultures with high PO and high AS may be influenced to use LM less superficially. By leveraging their cultural motivations, managers could encourage and facilitate a less-compromised LM activity by using strong LM champions, by setting aggressive and demanding performance targets, and by rewarding individuals for group performance. Finally, cultures with low UA by definition are less concerned or nervous about uncertainty, so require a more pronounced sense of urgency to take action. With low-UA employees, emphasis on the magnitude of damages possible from lack of performance or highlighting the required level (or reference point) of success may inspire a more regular application of LM. Using these kinds of insights will better prepare managers in designing how and where to implement LM practices and systems.

QUESTIONS TO CONSIDER

1. Does your organization view the world in an organic and ever-changing perspective, or does it tend to see it as and act as if it was a static model?
2. When considering the effectiveness of LM, do you make assumptions about which countries or cultures will be the most successful?
3. Are operational performance criteria or productivity considerations used when determining locations for new (foreign) facilities, or are financial criteria the sole determinant?
4. How does your organization adjust its LM practices in countries whose cultures are inconsistent with LM values?
5. Are your modifications to LM implementation driven by *what* is deployed or not deployed, or do you also consider *how* to deploy elements that may be counter to the local cultural norms?
6. Do you clearly define the potential cultural constraints and strategies to overcome them when launching new facilities?
7. What insights from other facilities do you use when designing LM deployment?
8. What cultural incongruities are more obvious than others? Can your management pinpoint specific mindsets that are hurting/helping the LM initiative?

7

Conclusions

THE GLOBAL PROFILE

Lean manufacturing (LM) is no longer a Japanese system introduced to the United States and other "developed" Western countries. It is a global manufacturing phenomenon becoming engrained in the psyche of worldwide manufacturing. This global footprint points to the need to more fully understand the way in which varying cultures around the world respond to the underlying values of LM. Throughout this book, we have been investigating and analyzing the way in which global cultural values contribute to or detract from the effectiveness of LM and its impact on operational performance.

To arrive at meaningful observations and conclusions, it is necessary to examine academic and professional literature and existing research on global cultural values and biases and develop the potential linkage between culture and LM. Ultimately, however, the most important element is to apply these findings in the world of manufacturing. Having meaning in the classroom or academic journals is somewhat futile if there is no carryover into the real world of manufacturing performance. This final chapter dives further into this real-world application, creating a more "usable" management tool to facilitate a more successful implementation of LM.

THE GLOBAL CULTURES

Lean manufacturing managers from around the world who are charged with the introduction and implementation of LM systems possess a powerful tool for success when they fully understand the measurable effects of culture. Until now, however, there has been little, if any, description of what the cultural effects are, or even a clear description of the culture of specific countries around the world. The following provides exactly that information. While the cultural descriptors may seem archaic or even politically incorrect, they are those identified in the GLOBE (Global Leadership and Organizational Behavior Effectiveness) study, so they serve as the basis for our discussion.

To fully understand the various global regions/cultures, it is important to revisit the relative effect of each of the cultural values. Recall that uncertainty avoidance (UA) has a relatively weak effect on LM effectiveness; assertiveness (AS) and future orientation (FO) have a mild effect; and performance orientation (PO) has a strong effect. We now review the specific and overall implications of these effects for each cultural region of the world, using the general tendencies of regions for each cultural dimension that influences LM effectiveness. The regions and related countries are listed, followed by the general dimensional tendency and what it will mean to managers looking to make LM effective. The regional findings should be viewed with these effect factors in mind. Figure 7.1 shows the lean manufacturing effectiveness for the various countries evaluated throughout this study.

Region/Culture: Anglo

Countries (GLOBE): Australia, Canada, England, Ireland, New Zealand, South Africa (white sample); United States.

Uncertainty Avoidance: Low UA values. As a result, workers tend to be unresponsive to signals and dislike structure and detailed routines.

Assertiveness: These countries exhibit mid-level cultural AS. Employees display some struggles with cooperative use of equipment, materials, tools, or systems (that is, sharing) and difficulty with systemic and procedural improvement.

Future Orientation: There is also a mid-level distribution in this value, which translates to challenges in accepting a sense of urgency, to pursue productivity, and with rapid and immediate change.

FIGURE 7.1
This figure indicates the average LM effectiveness for each country in the study.

Performance Orientation: Mid-level characteristic that manifests in worker difficulties in cooperation, standardized behavior, and use of root cause analysis.

Overall: **Some** cultural values that result in negative effect on LM effectiveness and operational performance.

Region/Culture: Latin Europe

Countries (GLOBE): France, Israel, Italy, Portugal, Spain, Switzerland (French speaking).

Uncertainty Avoidance: Mid-level effect resulting in some lack of attention to manufacturing problem signals (andon, etc.) and difficulties in implementing/maintaining detailed routines.

Assertiveness: Mid-level cultural AS. Employees display some struggles with cooperative use of equipment, materials, tools, or systems (that is, sharing) and difficulty with systemic and procedural improvement.

Future Orientation: Mid-level distribution in this value, which translates to challenges in accepting a sense of urgency, to pursue productivity, and with rapid and immediate change.

Performance Orientation: Mid-level characteristic that manifests in worker difficulties in cooperation, standardized behavior, and use of root cause analysis.

Overall: **Some** cultural values will be evident that result in negative effect on LM effectiveness and operational performance.

Region/Culture: Nordic Europe

Countries (GLOBE): Denmark, Finland, Sweden.

Uncertainty Avoidance: Low UA values. As a result, workers tend to be unresponsive to signals and dislike structure and detailed routines.

Assertiveness: Mid-level cultural AS. Employees display some struggles with cooperative use of equipment, materials, tools, or systems (that is, sharing) and difficulty with systemic and procedural improvement.

Future Orientation: Low FO provides a positive impact on LM effectiveness. A sense of urgency does exist in the manufacturing operations relative to productivity improvements, stepwise plan deployment, and quick meetings to address issues.

Performance Orientation: Mid-level characteristic that manifests in worker difficulties in cooperation, standardized behavior, and use of root cause analysis.

Overall: **Some** cultural values will be evident that result in negative effect on LM effectiveness and operational performance.

Region/Culture: Germanic Europe

Countries (GLOBE): Austria, Germany, Netherlands, Switzerland.

Uncertainty Avoidance: Low UA values. As a result, workers tend to be unresponsive to signals and dislike structure and detailed routines.

Assertiveness: Low level of cultural AS. Employees enjoy knowledge sharing and embrace ideals. System effects are blamed or credited for operational performance (or lack of performance). Manufacturing and system procedures are followed.

Future Orientation: Low FO provides a positive impact on LM effectiveness. A sense of urgency does exist in the manufacturing operations relative to productivity improvements, stepwise plan deployment, and quick meetings to address issues.

Performance Orientation: Mid-level characteristic that manifests in worker difficulties in cooperation, standardized behavior, and use of root cause analysis.

Overall: **Few** cultural values will be evident that result in negative effects on LM effectiveness and operational performance.

Region/Culture: Eastern Europe

Countries (GLOBE): Albania, Georgia, Greece, Hungary, Kazakhstan, Poland, Russia, Slovenia.

Uncertainty Avoidance: High UA. Manufacturing workers are responsive to process signals, value structured improvement, and are attracted to detailed work routines.

Assertiveness: Mid level of cultural AS. Employees display some struggles with cooperative use of equipment, materials, tools, or systems (that is, sharing) and difficulty with systemic and procedural improvement.

Future Orientation: Mid-level value, which translates to challenges in accepting a sense of urgency, to pursue productivity, and problems with rapid and immediate change.

Performance Orientation: Mid-level characteristic that manifests in worker difficulties in cooperation, standardized behavior, and use of root cause analysis.

Overall: **Some** cultural values will be evident that result in negative effect on LM effectiveness and operational performance.

Region/Culture: Latin America

Countries (GLOBE): Argentina, Bolivia, Brazil, Colombia, Costa Rica, Ecuador, El Salvador, Guatemala, Mexico, Venezuela.

Uncertainty Avoidance: High UA. Manufacturing workers are responsive to process signals, value structured improvement, and are attracted to detailed work routines.

Assertiveness: Mid level of cultural AS. Employees display some struggles with cooperative use of equipment, materials, tools, or systems (that is, sharing) and difficulty with systemic and procedural improvement.

Future Orientation: High FO. Lack urgency in planning and execution. Look for the development of lengthy plans through extensive meetings and deep-dive analysis. Productivity is viewed as unmotivating.

Performance Orientation: High-level characteristic creating significant struggles with cooperation and reward recognition. Workers tend to avoid problem exposure and the pursuit of root cause analysis. Performance and manufacturing process issues are overly personalized.

Overall: **Many** cultural values will be evident that result in negative effect on LM effectiveness and operational performance.

Region/Culture: Sub-Saharan Africa

Countries (GLOBE): Namibia, Nigeria, South Africa (black sample), Zambia, Zimbabwe.

Uncertainty Avoidance: High UA. Manufacturing workers are responsive to process signals, value structured improvement, and are attracted to detailed work routines.

Assertiveness: Mid level of cultural AS. Employees display some struggles with cooperative use of equipment, materials, tools, or systems (that is, sharing) and difficulty with systemic and procedural improvement.

Future Orientation: High FO. Lack urgency in planning and execution. Look for the development of lengthy plans through extensive meetings and deep-dive analysis. Productivity is viewed as unmotivating.

Performance Orientation: Mid-level characteristic that manifests in worker difficulties in cooperation, standardized behavior, and use of root cause analysis.

Overall: **Many** cultural values will be evident that result in negative effect on LM effectiveness and operational performance.

Region/Culture: Middle East

Countries (GLOBE): Egypt, Kuwait, Morocco, Qatar, Turkey.

Uncertainty Avoidance: High UA. Manufacturing workers are responsive to process signals, value structured improvement, and are attracted to detailed work routines.

Assertiveness: Low level of cultural AS. Employees enjoy knowledge sharing and embrace ideals. System effects are blamed or credited for operational performance (or lack of performance). Manufacturing and system procedures are followed.

Future Orientation: High FO. Lack urgency in planning and execution. Look for the development of lengthy plans through extensive meetings and deep-dive analysis. Productivity is viewed as unmotivating.

Performance Orientation: Mid-level characteristic that manifests in worker difficulties in cooperation, standardized behavior, and use of root cause analysis.

Overall: **Some** cultural values will be evident that result in negative effect on LM effectiveness and operational performance.

Region/Culture: Southern Asia

Countries (GLOBE): India, Indonesia, Iran, Malaysia, Philippines, Thailand.

Uncertainty Avoidance: High UA. Manufacturing workers are responsive to process signals, value structured improvement, and are attracted to detailed work routines.

Assertiveness: High level of cultural AS. Employees hold knowledge and tend to be against information sharing. Opportunism takes priority over and hurts system ideals. People and not systems are blamed for performance problems. Work procedures are not followed.

Future Orientation: High FO. Lack urgency in planning and execution. Look for the development of lengthy plans through extensive meetings and deep-dive analysis. Productivity is viewed as unmotivating.

Performance Orientation: Mid-level characteristic that manifests in worker difficulties in cooperation, standardized behavior, and use of root cause analysis.

Overall: **Many** cultural values will be evident that result in negative effect on LM effectiveness and operational performance.

Region/Culture: Confucian Asia

Countries (GLOBE): China, Hong Kong, Japan, Singapore, South Korea, Taiwan.

Uncertainty Avoidance: Mid level of cultural AS. Employees display some struggles with reacting to problem signal and adhering to detailed work routines.

Assertiveness: High level of cultural AS. Employees hold knowledge and tend to be against information sharing. Opportunism takes priority over and hurts system ideals. People and not systems are blamed for performance problems. Work procedures are not followed.

Future Orientation: Mid-level value, which translates to challenges in accepting a sense of urgency, to pursue productivity, and problems with rapid and immediate change.

Performance Orientation: Low-level characteristic where workers enjoy cooperation and revealing problems. Group rewards are motivating.

Overall: **Some** cultural values will be evident that result in negative effect on LM effectiveness and operational performance.

RANKING THE CULTURES AND COUNTRIES

A hierarchy of compatible cultures can be extrapolated from the previous analysis and table. Those identified with "few problems" are those countries that tend to be the most supportive or hospitable to LM values. Those with "many issues" reflect countries where implementation and success of LM would be difficult.

Few Issues: Austria, Germany, Netherlands, Switzerland

Some Issues: Albania, Australia, Canada, China, Denmark, Egypt, England, Finland, France, Georgia, Greece, Hong Kong, Hungary, Ireland, Israel, Italy, Japan, Kazakhstan, Kuwait, Morocco, New Zealand, Poland, Portugal, Qatar, Russia, Singapore, Slovenia, Spain, South Africa (white sample), South Korea, Sweden, Switzerland (French speaking), Turkey, Taiwan, United States

Many Issues: Argentina, Bolivia, Brazil, Colombia, Costa Rica, Ecuador, El Salvador, Guatemala, India, Indonesia, Iran, Malaysia, Mexico, Namibia, Nigeria, Philippines, South Africa (black sample), Zambia, Thailand, Venezuela, Zimbabwe

The study analysis and hypothesis testing completed in the research by Kull, Yan and Wacker (2010), and as the basis for this book, also showed measurable influence on LM effectiveness. The chart in Table 7.1 indicates the "average LM effectiveness" for a selected cross section of countries.

TABLE 7.1

Regional LM Cultural Implications

		National Cultural Value Dimensions Relevant to Lean Manufacturing				
Region	**Countries in GLOBE Study**	**Uncertainty Avoidance (Weak Effect)**	**Assertiveness (Mild Effect)**	**Future Orientation (Mild Effect)**	**Performance Orientation (Strong Effect)**	**Overall**
Anglo	Australia, Canada, England, Ireland, New Zealand, South Africa (white sample), United States	**Low**: Unresponsive to signals, dislike structure and detailed routines	**Mid**: Some struggles with sharing and systemic, procedural improvement	**Mid**: Some struggles with urgency, productivity, and rapid change	**Mid**: Some struggles with cooperative, standardized behavior and root cause analysis	Some cultural problems
LM effect		↓	↘	↘	↘	↘
Latin Europe	France, Israel, Italy, Portugal, Spain, Switzerland (French speaking)	**Mid**: Some struggles with problem signals and detailed routines	**Mid**: Some struggles with sharing and systemic, procedural improvement	**Mid**: Some struggles with urgency, productivity, and rapid change	**Mid**: Some struggles with cooperative, standardized behavior and root cause analysis	Some cultural problems
LM effect		↘	↘	↘	↘	↘

continued

TABLE 7.1 (continued)

Regional LM Cultural Implications

			National Cultural Value Dimensions Relevant to Lean Manufacturing			
Region	**Countries in GLOBE Study**	**Uncertainty Avoidance (Weak Effect)**	**Assertiveness (Mild Effect)**	**Future Orientation (Mild Effect)**	**Performance Orientation (Strong Effect)**	**Overall**
Nordic Europe	Denmark, Finland, Sweden	**Low**: Unresponsive to signals, dislike structure and detailed routines	**Mid**: Some struggles with sharing and systemic, procedural improvement	**Low**: Urgency exists to productivity and stepwise plans, quick meetings	**Mid**: Some struggles with cooperative, standardized behavior and root cause analysis	Some cultural problems
LM effect		↓	↘	↑	↘	↘
Germanic Europe	Austria, Germany, Netherlands, Switzerland	**Low**: Unresponsive to signals, dislike structure and detailed routines	**Low**: Enjoy knowledge sharing and ideals, system is blamed, procedures are followed	**Low**: Urgency exists to productivity and stepwise plans, quick meetings	**Mid**: Some struggles with cooperative, standardized behavior and root cause analysis	Few cultural problems
LM effect		↓	↑	↑	↘	↗

Eastern Europe	Albania, Georgia, Greece, Hungary, Kazakhstan, Poland, Russia, Slovenia	**High**: Responsive to signals, value structured improvement, attracted to detailed routines	**Mid**: Some struggles with sharing and systemic, procedural improvement	**Mid**: Some struggles with urgency, productivity, and rapid change	**Mid**: Some struggles with cooperative, standardized behavior and root cause analysis	Some cultural problems
LM effect		↑	↘	↘	↘	↘
Latin America	Argentina, Bolivia, Brazil, Colombia, Costa Rica, Ecuador, El Salvador, Guatemala, Mexico, Venezuela	**High**: Responsive to signals, value structured improvement, attracted to detailed routines	**Mid**: Some struggles with sharing and systemic, procedural improvement	**High:** Lacks urgency, lengthy plans, extensive meetings, deep analyses, productivity unmotivating	**High**: Struggle with cooperation and no rewards, avoid problem exposure and root causes, overly personalized	Many cultural problems
LM effect		↑	↘	↓	↓	↓
Sub-Saharan Africa	Namibia, Nigeria, South Africa (black sample), Zambia, Zimbabwe	**High:** Responsive to signals, value structured improvement, attracted to detailed routines	**Mid:** Some struggles with sharing and systemic, procedural improvement	**High:** Lacks urgency, lengthy plans, extensive meetings, deep analyses, productivity unmotivating	**Mid:** Some struggles with cooperative, standardized behavior and root cause analysis	Many cultural problems
LM effect		↑	↘	↓	↘	↓

continued

TABLE 7.1 (continued)

Regional LM Cultural Implications

		National Cultural Value Dimensions Relevant to Lean Manufacturing				
Region	**Countries in GLOBE Study**	**Uncertainty Avoidance (Weak Effect)**	**Assertiveness (Mild Effect)**	**Future Orientation (Mild Effect)**	**Performance Orientation (Strong Effect)**	**Overall**
Middle East	Egypt, Kuwait, Morocco, Qatar, Turkey	**High:** Responsive to signals, value structured improvement, attracted to detailed routines	**Low:** Enjoy knowledge sharing and ideals, system is blamed, procedures are followed	**High:** Lacks urgency, lengthy plans, extensive meetings, deep analyses, productivity unmotivating	**Mid:** Some struggles with cooperative, standardized behavior and root cause analysis	Some cultural problems
LM effect		↑	↑	↓	↘	→
Southern Asia	India, Indonesia, Iran Malaysia, Philippines, Thailand	**High:** Responsive to signals, value structured improvement, attracted to detailed routines	**High:** Against knowledge sharing, opportunism hurts system ideals, people are blamed, procedures are not followed	**High:** Lacks urgency, lengthy plans, extensive meetings, deep analyses, productivity unmotivating	**Mid:** Some struggles with cooperative, standardized behavior and root cause analysis	Many cultural problems
LM effect		↑	↓	↓	↘	↓

Confucian LM effect		↘	↓	↘	↑	→
Asia	China, Hong Kong, Japan, Singapore, South Korea, Taiwan	**Mid:** Some struggles with problem signals and detailed routines	**High:** Against knowledge sharing, opportunism hurts system ideals, people are blamed, procedures are not followed	**Mid:** Some struggles with urgency, productivity, and rapid change	**Low:** Enjoy cooperation and revealing problems, group rewards are motivating	Some cultural problems

This distribution reflects not only the average LM effectiveness for some of the countries in the study but also the relative relationship between those countries. Many of the Western industrialized nations, including the United States, demonstrated a relatively low level of LM effectiveness. Many Eastern European and Asian-Pacific nations showed the highest effectiveness. It can be interpreted from the data that countries like Croatia, Korea, Albania, Macedonia, and Taiwan have country cultures that are more aligned with the key underlying Lean values than countries such as New Zealand, Italy, Canada, and the United States.

SUMMARY

Many of the recent studies of LM emphasize that its success lies in the culture of the people responsible for the implementation and day-to-day utilization of LM practices. These studies have largely been limited to descriptive or anecdotal accounts of practitioners or observers of lean systems. We attempted to move beyond these descriptive accounts by articulating the specific values underpinning LM and then showing those values empirically important for *predicting* LM effectiveness. It is these evidence-based value results that provide a deeper understanding of why certain countries are more successful in implementing and using LM practices and why others have more difficulty.

We have focused our research discussion on country-level effectiveness. As we hypothesized, country culture is found to be significantly influential; however, the importance of organizational culture cannot be ignored. Just as Toyota has cultural characteristics dissimilar to Japanese culture, LM managers should expect that each organization, even within the same company and within the same country, will have its own unique features. Future investigation and research should focus on the ways in which organizational culture interacts with country culture and how LM is affected. This future research should continue to utilize global research data sets, such as the GMRG (Global Manufacturing Research Group) and GLOBE databases that we used. In this way, future empirical testing and theory development/refinement will help build a clearer picture of the interaction of culture and LM.

QUESTIONS TO CONSIDER

1. Do you see the countries of your company's expansion among the analyses shown here? How would you begin to accommodate the cultural variations in your country specific Lean deployment?
2. What preconceived notions regarding culture do you find validated or contrary to what you had expected?
3. Would the information provided in the country culture–Lean effectiveness analysis make you reconsider the countries where you might expand your operations?
4. How would you begin to consider or separate the effect of country culture on LM effectiveness versus company culture, such as Japanese cultural values/effects versus Toyota company culture/effects?
5. With the cultural effects on LM effectiveness more clearly understood, will you and your organization be able to utilize Lean "tools" and practices more effectively? How?

Appendix: Studies, Data, and Analysis

LEAN EXPERT SURVEY

Because some of the literature-based hypotheses were unsupported, we administered an online survey of individuals known to be experts in Lean manufacturing (LM). While beyond the scope of this book, we view these data as important for students, managers, and researchers looking to continually improve their understanding of the values of LM. By reading the results given in this appendix, you will refine your knowledge regarding what underlies the thinking of LM.

Nineteen LM experts agreed to complete the survey, with sixteen completing it in full. The experts had an average of twenty-eight years of work experience and sixteen years of experience working with LM systems. Seven experts primarily had experience with LM in the automotive industry. Other industry experiences noted were electronics; aerospace; process (glass, paper, steel, refining, sputtering); light truck supply; wind turbines; HVAC (heating, ventilation, air-conditioning); capital equipment manufacturing; plastic injection molding; polyurethane foam molding; transportation services and equipment; legal text and case law digests; consumer food and other products; yacht manufacturing; consulting; semiconductor and other packaging; appliances; displays; business services; and higher education.

Some specific experiences noted with respect to LM use were as follows:

- "Led first lean transformation in 1. Automotive interior supply, 2. Heavy duty trucks, Sputtering technology, 3. Wind turbine generators."
- "In Transportation primarily in the sourcing and manufacturing areas. Most specifically in work cell setups and in production line optimization."
- "Executive leadership for the entire time. 5 years consulting."

- "Consulting."
- "High tech electronic fabrication and assembly. Industrial manufacturing."
- "Consulting, speaking."
- "All facets from consulting with plants on the training and 'how to implement' to being the plant manager and overseeing the implementation itself."
- "Light assembly."
- "I have been involved in taking Lean Approaches at Corporate HQ and R&D [research and development]."
- "Automotive assembly, other light assembly, and consulting."
- "Flexible automation."
- "General management."

Lean Manufacturing Definition

Our definition of LM based on Shah and Ward (2007) was "Lean Manufacturing (LM) is here defined as a set of inter-related components that focuses on (1) attaining continuous production flow, (2) having total preventative maintenance, (3) allowing employee involvement, (4) using pull production system, (5) reducing setup times, and (6) utilizing statistical process control."

Fifteen of sixteen agreed or strongly agreed with this literature-based definition.

Comments regarding this definition from the experts were as follows:

- "Manufacturing manager's ownership and leadership should be included."
- "Line-side material storage, fail-safe measures to stop defect propagation, motivated people & trained teams, focus on customer satisfaction, a unified and integrated quality system."
- "My response is based more on LM 'training and education' more than on our specific implementations of LM principals."
- "I would stop at number 4 and add number 5 as In Station Process Control. I would take out reducing setup time and SPC [statistical process control] as these are both just tools enabling these above items to be accomplished."
- "Eliminates waste/cost, continuous improvement."
- "LM is engaging all employees at all levels in continuous improvement."

- "Statistical process control is not a part of the TPS [Toyota Production System] system as it was taught to me but rather part of SPC. In addition, the Preventative Maintenance aspect is normally not emphasized (and it probably should be)."
- "I like to see continuous improvement added, but that can be a separate subject."
- "Employee involvement is strongly desired of course and results won't be as effective without it, but I would not include it in the above definition."
- "Just wondering if 'allowing employee involvement' covers all the OB type practices associated with LM."
- "I'm not sure if SPC is so essential. Some essential elements, such as reducing batch size and inventory, reducing lead time, and most of all, reducing WASTE, are missing."

The remainder of this report lists each of the eight dimensions by degree of distinction on the dimension's primary dichotomy. We give the overall score toward each dimension's primary dichotomy (higher means stronger agreement) and then the overall agreement/disagreement toward the literature-based LM value within that dimension. Comments are also provided.

VALUE DIMENSIONS WITH CLEAR DISTINCTIONS

Truth

Experts clearly chose *rationality* (score of 4.3) instead of *intuition* (score of 2.3).

Supporting experts stated: "LM relies on a set of concepts that are rather invariant." "many times intuition can lead you to the general area, but the rationality many times proves that intuition is not as accurate as we think it is or would like it to be." "My experience is LM requires measurement and things that can be measured (facts = truth). However, many times gut feel and intuition play a part to ensure you are improving things that really add value to the bottom line." "My belief is that there should be a (clear) balance, and Design Thinking principles can be applied to identifying opportunities and solving challenges in new ways. Reality is that most corporations focus on analysis and rationality." "Lean thinking is 'taught' to those that have not been exposed to it—hence it is not intuitive." "Rationality fits better with ideas of statistical process control, pull systems, etc., where the goal is to reduce variance in decision making and are using tools to make rational decisions."

Nonsupporting experts stated: Not applicable.

Twelve of sixteen agreed or strongly agreed with the literature-based LM truth value: "LM values decisions based on visible, first-hand facts and problems in a world that's difficult to predict."

Supporting experts stated: Empirical data.

Nonsupporting experts stated: "The main task of LM is to reduce variability so that prediction is one of the central tenets." "Not sure of best wording, but this feels too restrictive."

Motivation

Experts clearly chose *intrinsic* (score of 3.9) instead of *extrinsic* (score of 2.7).

Supporting experts stated: "Once the system is running many of the engagements from the team members are intrinsic however much of the success is extrinsic from being a part of the larger group working together." "Purpose of including employees is to help drive intrinsic motivation to root out problems in their area of expertise."

Nonsupporting experts stated: "People basically respond to challenge and are motivated if they are treated fairly and given a chance to improve both the organization and themselves at the same time." "Production workers probably need extrinsic motivation to see and implement LM." "I believe balance is critical. I subscribe to the belief that once basic needs are met, people will be motivated by opportunities to make a meaningful difference." "People in industry generally do not change on their own but rather wait to have change implemented." "The standard 'if it is not broke don't fix it' mentality."

Twelve of sixteen agreed or strongly agreed with the literature-based LM motivation value: "LM values a view that problems are systemic and helps reveal issues that people intrinsically are challenged to solve."

Supporting experts stated: Not applicable.

Nonsupporting experts stated: "I am just not too sure on this one." "Keep in mind that TPS is essentially a military system one-fourth of people are told what to do and there are rewards (generally) if they do it. Externally driven."

Vocation

Experts clearly chose *productivity* (score of 4.0) instead of *relationships* (score of 3.1).

Supporting experts stated: "In my company, we tend to focus on products and productivity in building those products. However, my personal experience is that it takes people (and therefore relationships) to really make a difference. However, productivity is the measurement and therefore why I rated it slightly higher." "Both are important, but ultimate focus on results takes precedence." "TPS is task oriented and group compliant oriented." "I don't see these as mutually exclusive." "Both are required in practice."

Nonsupporting experts stated: "LM should focus on simultaneous improvements in both people's lives and the organization."

Twelve of sixteen agreed or strongly agreed with the literature-based LM vocation value: "LM values that productivity needs continuous monitoring with distinct work duties among operators and leaders."

Supporting experts stated: "Hour by hour charts are a driving force one-fourth what happened, what should have happened, how do we fix it?"

Nonsupporting experts stated: "Very fuzzy statement. Quit using the word productivity. It has too much baggage." "LM values the use of standardization to facilitate continuous improvement and the work duties often change for the operators and leaders however productivity is maintained and improved by the team." "Change productivity to performance."

Interdependence

Experts clearly chose *collaboration* (score of 4.3) instead of *isolation* (score of 1.8).

Supporting experts stated: "Focus on teams." "Handoffs within the production environment are critical with mass produced products like most of ours." "Both are important. In order to optimize synergies, each has to be the best they can be in their own functional area for the interdependence to work."
Nonsupporting experts stated: Not applicable.

Fourteen of sixteen agreed or strongly agreed with the literature-based LM interdependence value: "LM values that the entire workforce is interdependent and must cooperate using guidelines to solve problems."

Supporting experts stated: "I might add something about individual excellence." "Manufacturing is a team sport and all parts of it are driven to help production do 'its thing.'"
Nonsupporting experts stated: Not applicable.

VALUE DIMENSIONS WITH A SLIGHT DISTINCTION

Time

Experts somewhat chose *long-term* (score of 3.6) instead of *short-term* (score of 2.8).

Supporting experts stated: "Must be committed for the long-term but implementation should have a very short-term focus." "Some flexibility required for real world situations." "Both are required in practice."
Nonsupporting experts stated: "Often times, the biggest gains are simple and can be understood and implemented quickly with little training. Example is new technology that revolutionizes and improves effectiveness. My field teams all have iPads and we have a proprietary iPad app. Now all sales and service information is available at all times and at any location." "Most lean practices are waste reduction from short-term issues and small waste identification and elimination however that must be solved and a new standard set to ensure root cause results and long term resolution." "I believe balance is crucial to Lean success, short-, medium-, and long-term—aligned with corporate strategy and multiyear plans." "The value stream map shows three states: present, 6 months from now, ideal. This suggests short-, intermediate-, and long-term thinking."

Fifteen of sixteen agreed or strongly agreed with the literature-based LM time value: "LM values urgent timeliness toward production and quick improvements guided by long-term ideals."

Supporting experts stated: "Well said." "If 'production' is the focus, this is fine, but we see companies applying Lean to areas beyond manufacturing and production." "Agreed one-fourth and this also plays into the whole TPM discussion and how it tends to lag behind the other components of TPS."
Nonsupporting experts stated: Not applicable.

Control

Experts somewhat chose *decentralized* (score of 3.5) instead of *centralized* (score of 2.8).

Supporting experts stated: Not applicable.
Nonsupporting experts stated: "Basic concepts should be developed for all. Tools are offered for assistance in implementing." "Centralized in that a process for implementing LM is required or the results will not materialize." "Centralized for consistency across the organization." "It is a balanced approach." "TPS is driven from the top. Now once established, improvements can be made at the group level, but the primary reporting mechanism is still driven from the top one-fourth the institutionalization of TPS." "Depends on what is 'decentralized.'" "Company structure would impact for me." "Not sure I see it one way or the other. On one hand, many improvements seem centrally controlled such as statistical process control. However, employee involvement also elicits decentralized decision making."

Thirteen of sixteen agreed or strongly agreed with the literature-based LM control value: "LM values detailed routines to create standards that temper irrationality and uncertainty while channeling human capabilities."

Supporting experts stated: "Decentralized elements should emphasize trust in individuals to be creative and allow decision making, supported by tools, processes, investment, etc."
Nonsupporting experts stated: "Standards that are continually improved."

STUDY DATA

TABLE A1

Sample Demographics

	Percentage
Number of facility employees	
<50	23.9
50–100	24.8
101–500	37.2
>500	14.1
Number of product lines	
<5	48.7
5–10	33.3
>10	18.0
Average age of equipment	
<5 years	14.0
5–10 years	48.4
>10 years	38.6
Industrial groups in sample	
Computing, electronics, and related products	28.4
Food and related products	7.8
Metal-based products	19.1
Petrochemicals and related products	8.3
Textiles, clothing, and footwear	5.1
Timber and related products	8.3
Miscellaneous or not reported	23.1
Mean percentage sales from exports	34.1
Mean percentage materials imported	34.8
Mean percentage international ownership	24.0
Mean percentage of machines grouped by	
Type	41.5
Product	36.4
Assembly line	26.1

TABLE A2

Rotated Component Matrix[a]

	Component			
	1	2	3	4
Performance compared to competitors on				
Manufacturing costs	.175	.172	.887	.157
Product costs	.166	.184	.896	.116
Product performance	.163	.148	.146	.879
Product quality	.120	.272	.122	.853
Order fulfillment speed	.135	.872	.154	.133
Delivery speed	.162	.900	.138	.129
Delivery as promised	.144	.839	.116	.208
Extent invested resources in				
Cellular manufacturing	.605	−.031	.136	.148
Process redesign	.642	−.039	.097	.096
JIT	.678	.214	.072	.045
Manufacturing throughput time reduction	.779	.107	.028	.111
Setup time reduction	.780	.142	.093	.040
Statistical process control	.593	.156	−.002	.096
Waste reduction	.555	.137	.156	−.039

[a] Extraction method: principal component analysis. Rotation method: varimax with Kaiser normalization.

TABLE A.2

[illegible]

Bibliography

Abrahamson, E., 1991. Managerial Fads and Fashions—The Diffusion and Rejection of Innovations. *Academy of Management Review* 16 (3), 586–612.

Ahmad, S., Schroeder, R.G., and Sinha, K.K., 2003. The Role of Infrastructure Practices in the Effectiveness of JIT Practices: Implications for Plant Competitiveness. *Journal of Engineering and Technology Management* 20 (3), 161–191.

Ansari, S.M., Fiss, P.C., and Zajac, E.J., 2010. Made to Fit: How Practices Vary as They Diffuse. *Academy of Management Review* 35, 67–92.

Aoki, K., 2008. Transferring Japanese Kaizen Activities to Overseas Plants in China. *International Journal of Operations & Production Management* 28 (6), 518–539.

Beale, D., 1994. *Driven by Nissan? A Critical Guide to New Management Techniques*. London: Lawrence and Wishart.

Best, D.L., Williams, J.E., Cloud, J.M., Davis, S.W., Robertson, L.S., Edwards, J.R., Giles, H., and Fowles, J., 1977. Development of Sex-Trait Stereotypes among Young Children in the United States, England, and Ireland. *Child Development* 48, 1375–1384.

Boulton, J., 2010. Complexity Theory and Implications for Policy Development. *Emergence: Complexity and Organization* 12, 31–40.

Brown, S. R., 1996. Q Methodology and Qualitative Research. *Qualitative Health Research* 6 (4) 261–267.

Browning, T.R., and Health, R.D., 2009. Reconceptualizing the Effects of Lean on Production Costs with Evidence from the F-22 Program. *Journal of Operations Management* 27 (1), 23–44.

Cai, S., Jun, M., and Yang, Z., 2010. Implementing Supply Chain Information Integration in China: The Role of Institutional Forces and Trust. *Journal of Operations Management* 28 (3), 257–268.

Carl, D., Gupta, V., and Javidan, M., 2004. Power Distance. In: R.J. House, P.J. Hanges, M. Javidan, P.W. Dorfman, and V. Gupta (Eds.), *Leadership, Culture and Organization: The Globe Study of 62 Societies*, Sage, Thousand Oaks, CA, pp. 512–563.

Cheng, P.C.H., and Dawson, S.D., 1998. A Study of Statistical Process Control: Practice, Problems and Training Needs. *Total Quality Management* 9 (1), 3–20.

Choi, T.Y., and Liker, J.K., 1995. Bringing Japanese Continuous Improvement Approaches to U.S. Manufacturing: The Roles of Process Orientation and Communications. *Decision Sciences* 26, 589–620.

Conti, R., Angelis, J., Cooper, C., Faragher, B., and Gill, C., 2006. The Effects of Lean Production on Worker Job Stress. *International Journal of Operations & Production Management* 26 (9), 1013.

Cua, K.O., McKone, K.E., and Schroeder, R.G., 2001. Relationships between Implementation of TQM, JIT, and TPM and Manufacturing Performance. *Journal of Operations Management* 19 (6), 675–694.

Dankbaar, B., 1997. Lean Production: Denial, Confirmation or Extension of Sociotechnical Systems Design? *Human Relations* 50 (5), 567–584.

Detert, J., R., Schroeder, R.G., and Mauriel, J.J., 2000. A Framework for Linking Culture and Improvement Initiatives in Organizations. Academy of Management. *Academy of Management Review* 25 (4), 850.

Diaby, M., 1995. Optimal Setup Time Reduction for a Single Product with Dynamic Demands. *European Journal of Operational Research* 85 (3), 532–540.

Douglas, T.J., and Fredendall, L.D., 2004. Evaluating the Deming Management Model of Total Quality in Services. *Decision Sciences* 35, 393–422.

Eade, R., 1995. *Cellular Manufacturing in a Global Marketplace—Video Reference Supplement.* Dearborn, MI: Society of Manufacturing Engineers.

Edström, A., and Olhager, J., 1987. Production-Economic Aspects on Set-up Efficiency. *Engineering Costs and Production Economics* 12 (1–4), 99–106.

Egbelu, P.J., and Wang, H.P., 1989. Scheduling for Just-in-Time Manufacturing. *Engineering Costs and Production Economics* 16 (2), 117–124.

Euwema, M.C., Wendt, H., and Van Emmerik, H., 2007. Leadership Styles and Group Organizational Citizenship Behavior across Cultures. *Journal of Organizational Behavior* 28, 1035–1057.

Fey, C.F., Morgulis-Yakushev, S., Park, H.J., and Björkman, I., 2009. Opening the Black Box of the Relationship between HRM Practices and Firm Performance: A Comparison of MNE Subsidiaries in the USA, Finland, and Russia. *Journal of International Business Studies* 40, 690–712.

Flynn, B.B., Sakakibara, S., and Schroeder, R.G., 1995. Relationship between JIT and TQM: Practices and Performance. *Academy of Management Journal* 38 (5), 1325.

Flynn, B.B., and Saladin, B., 2006. Relevance of Baldrige Constructs in an International Context: A Study of National Culture. *Journal of Operations Management* 24, 583–603.

Flynn, B.B., Schroeder, R.G., and Sakakibara, S., 1994. A Framework for Quality Management Research and an Associated Measurement Instrument. *Journal of Operations Management* 11 (4), 339–366.

Forza, D., 1996. Work Organization in Lean Production and Traditional Plants: What Are the Differences? *International Journal of Operations & Production Management* 16, 42–62.

Fucini, J., and Fucini, S., 1990. *Working for the Japanese.* New York: Free Press.

Gelfand, M.J., Bhawuk, D.P.S., Nishii, L.H., and Bechtold, D.J., 2004. Individualism and Collectivism. In: R.J. House, P.J. Hanges, M. Javidan, P.W. Dorfman, and V. Gupta (Eds.), *Leadership, Culture and Organization: The Globe Study of 62 Societies*, Sage, London.

Gotoh, F., 1991. *Equipment Planning for TPM: Maintenance Prevention Design.* Cambridge, UK: Productivity Press.

Gottfried, H., and Graham, L., 1993. Constructing Difference: The Making of Gendered Subcultures in a Japanese Automobile Assembly Plant. *Sociology* 27 (4), 611–628.

Graham, J.L., 1985. The Influence of Culture on the Process of Business Negotiations: An Exploratory Study. *Journal of International Business Studies* 16, 81–96.

Green, S.D., 1999. The Missing Arguments of Lean Construction. *Construction Management and Economics* 17 (2), 133–137.

Gupta, V., and Hanges, P.J., 2004. Regional and Climate Clustering of Societal Cultures. In: R.J. House, P.J. Hanges, M. Javidan, P.W. Dorfman, and V. Gupta (Eds.), *Leadership, Culture and Organization: The Globe Study of 62 Societies*, Sage, London.

Hahn, E.D., and Bunyaratavej, K., 2010. Services Cultural Alignment in Offshoring: The Impact of Cultural Dimensions on Offshoring Location Choices. *Journal of Operations Management* 28, 186–193.

Hale, J.R., and Fields, D.L., 2007. Exploring Servant Leadership across Cultures: A Study of Followers in Ghana and the USA. *Leadership* 3, 397–418.

Hanges, P.J., and Dickson, M.W., 2004. The Development and Validation of the Globe Culture and Leadership Scales. In: R.J. House, P.J. Hanges, M. Javidan, P.W. Dorfman, and V. Gupta (Eds.), *Leadership, Culture and Organization: The Globe Study of 62 Societies*, Sage, London.

Hartog, D.N.D., 2004. Assertiveness. In: R.J. House, P.J. Hanges, M. Javidan, P.W. Dorfman, and V. Gupta (Eds.), *Leadership, Culture and Organization: The Globe Study of 62 Societies*, Sage, London.

Hennart, J.-F., and Larimo, J., 1998. The Impact of Culture on the Strategy of Multinational Enterprises: Does National Origin Affect Ownership Decisions? *Journal of International Business Studies* 29, 515–538.

Higgins, E.T., 1997. Beyond Pleasure and Pain. *American Psychologist* 52, 1280–1300.

Hirano, H., 1988. *JIT Factory Revolution: A Pictorial Guide to Factory Design of the Future.* Portland, OR: Productivity Press.

Hofmann, D.A., 1997. An Overview of the Logic and Rationale of Hierarchical Linear Models. *Journal of Management* 23 (6), 723–744.

Hofstede, G., 1980. *Culture's Consequences: International Difference in Work-Related Values.* Beverly Hills, CA: Sage.

Hofstede, G., 2001. *Culture's Consequences: Comparing Values, Behaviors, Institutions and Organizations across Nations,* 2nd ed. Thousand Oaks, CA: Sage.

Hopp, W.J., and Spearman, M.L., 2004. To Pull or Not to Pull: What Is the Question? *Manufacturing and Service Operations Management* 6, 133.

House, R.J., Hanges, P.J., Javidan, M., Dorfman, P.W., and Gupta, V., Eds., 2004. *Culture, Leadership, and Organizations: The Globe Study of 62 Societies*. London: Sage.

Huber, V.L., and Brown, K.A., 1991. Human Resource Issues in Cellular Manufacturing: A Sociotechnical Analysis. *Journal of Operations Management* 10 (1), 138–159.

Hui, M.K., Au, K., and Fock, H., 2004. Empowerment Effects across Cultures. *Journal of International Business Studies* 35, 46–60.

Hyer, N.L., Brown, K.A., and Zimmerman, S., 1999. A Socio-Technical Systems Approach to Cell Design: Case Study and Analysis. *Journal of Operations Management* 17 (2), 179–203.

Imai, M., 1986. *Kaizen: The Key to Japan's Competitive Success.* New York: McGraw-Hill/Irwin.

Ishida, H., 1986. Transferability of Japanese Human-Resource Management Abroad. *Human Resource Management* 25, 103–120.

Jayaram, J., Das, A., and Nicolae, M., 2010. Looking Beyond the Obvious: Unraveling the Toyota Production System. *International Journal of Production Economics* 128, 280–291.

Javidan, M., 2004. Performance Orientation. In: R.J. House, P.J. Hanges, M. Javidan, P.W. Dorfman, and V. Gupta (Eds.), *Leadership, Culture and Organization: The Globe Study of 62 Societies*, Sage, London.

Kabasakal, H., and Bodur, M., 2004. Human Orientation in Societies, Organizations, and Leader Attributes. In: R.J. House, P.J. Hanges, M. Javidan, P.W. Dorfman, and V. Gupta (Eds.), *Leadership, Culture and Organization: The Globe Study of 62 Societies*, Sage, London.

Kearney, M., 1984. *World View.* Novato, CA: Chandler and Sharp.

Keough, K.A., Zimbardo, P.G., and Boyd, J.N., 1999. Who's Smoking, Drinking, and Using Drugs? Time Perspective as a Predictor of Substance Use. *Basic and Applied Social Psychology* 21, 149–164.

Khazanchi, S., Lewis, M.W., and Boyer, K.K., 2007. Innovation-Supportive Culture: The Impact of Organizational Values on Process Innovation. *Journal of Operations Management* 25, 871–884.

Khurana, A., 1999. Managing Complex Production Processes. *MIT Sloan Management Review* 40, 85–97.

Kirca, A.H., and Hult, G.T.M., 2009. Intra-Organizational Factors and Market Orientation: Effects of National Culture. *International Marketing Review* 26, 633–650.

Kluckhohn, F.R., and Strodtbeck, F.L., 1973. *Variations in Value Orientations*. Westport, CT: Greenwood Press.

Kogut, B., and Singh, H., 1988. The Effect of National Culture on the Choice of Entry Mode. *Journal of International Business Studies* 19, 411–432.

Koufteros, X.A., and Vonderembse, M.A., 1998. The Impact of Organizational Structure on the Level of JIT Attainment: Towards Theory Development. *International Journal of Production Research* 36 (10), 2863–2878.

Kull, T.J., and Wacker, J.G., 2010. Quality Management Effectiveness in Asia: The Influence of Culture. *Journal of Operations Management* 28 (3), 223–239.

Leggett, S., 2010. TPS Troubles. *Quality Progress* 43 (4), 8.

Liker, J., 1997. *Becoming Lean: Inside Stories of U.S. Manufacturers*. New York: Productivity Press.

Liker, J.K., 2003. *The Toyota Way*. New York: McGraw-Hill.

Liker, J., and Hoseus, M., 2010. Human Resource Development in Toyota Culture. *International Journal of Human Resources Development and Management* 10 (1), 34.

Liker, J.K., and Hoseus, M., 2008. *Toyota Culture: The Heart and Soul of the Toyota Way*. New York: McGraw-Hill.

Liker, J., and Rother, M., 2011. *Why Lean Programs Fail*. Cambridge, MA: Lean Enterprise Institute.

Lozeau, D., Langley, A., and Denis, J.-L., 2002. The Corruption of Managerial Techniques by Organizations. *Human Relations* 55 (5), 537–564.

Luo, Y.D., 2008. Procedural Fairness and Interfirm Cooperation in Strategic Alliances. *Strategic Management Journal* 29, 27–46.

MacDuffie, J.P., and Krafcik, J.F., 1992. Integrating Technology and Human Resources for High Performance Manufacturing: Evidence from the International Auto Industry. In: T.A. Kochan and M. Useem (Eds.), *Transforming Organizations*, Oxford University Press, New York.

The Machine that Ran Too Hot; Toyota's Overstretched Supply Chain, 2010. *Economist* 394 (8671), 74.

Mackelprang, A.W., and Nair, A., 2010. Relationship between Just-in-Time Manufacturing Practices and Performance: A Meta-Analytic Investigation. *Journal of Operations Management* 28, 283–302.

Mann, D., 2010. *Creating a Lean Culture: Tools to Sustain Lean Conversations*, 2nd ed. New York: Productivity Press.

McClelland, D.C., 1987. *Human Motivation*. Cambridge: Cambridge University Press.

McDermott, C.M., and Stock, G.N., 1999. Organizational Culture and Advanced Manufacturing Technology Implementation. *Journal of Operations Management* 17, 521–533.

McKone, K.E., Schroeder, R.G., and Cua, K.O., 1999. Total Productive Maintenance: A Contextual View. *Journal of Operations Management* 17 (2), 123–144.

McKone, K.E., Schroeder, R.G., and Cua, K.O., 2001. The Impact of Total Productive Maintenance Practices on Manufacturing Performance. *Journal of Operations Management* 19 (1), 39–58.

McKone, K.E., and Weiss, E., 1998. TPM: Planned and Autonomous Maintenance: Bridging the Gap between Practice and Research. *Production and Operations Management* 7 (4), 335–351.

McLachlin, R., 1997. Management Initiatives and Just-in-Time Manufacturing. *Journal of Operations Management* 15 (4), 271–292.

Meyer, K., 2007, February 7. The Culture Side of Lean Manufacturing. Evolving Excellence. http://www.evolvingexcellence.com/blog/2007/02/the_culture_sid.html#ixzz1EF4nQmqP, accessed February 17, 2011.

Miron, E., Erez, M., and Naveh, E., 2004. Do Personal Characteristics and Cultural Values That Promote Innovation, Quality, and Efficiency Compete or Complement Each Other? *Journal of Organizational Behavior* 25, 175–199.

Morris, T., and Pavett, C.M., 1992. Management Style and Productivity in Two Cultures. *Journal of International Business Studies* 23, 169–179.

Nahm, A.Y., Vonderembse, M.A., and Koufteros, X.A., 2004. The Impact of Organizational Culture on Time-Based Manufacturing and Performance. *Decision Sciences* 35 (4), 579–608.

Nakajima, S., 1988. *Introduction to TPM*. Cambridge, MA: Productivity Press.

Nakata, C., and Sivakumar, K., 1996. National Culture and New Product Development: An Integrative Review. *Journal of Marketing* 60 (1), 61–72.

Naor, M., Goldstein, S.M., Linderman, K.W., and Schroeder, R.G., 2008. The Role of Culture as Driver of Quality Management and Performance: Infrastructure versus Core Quality Practices. *Decision Sciences* 39, 671–702.

Naor, M., Linderman, K., and Schroeder, R.G., 2010. The Globalization of Operations in Eastern and Western Countries: Unpacking the Relationship between National and Organizational Culture and its Impact on Manufacturing Performance. *Journal of Operations Management* 28 (3), 194–205.

Newman, K.L., and Nollen, S.D., 1996. Culture and Congruence: The Fit between Management Practices and National Culture. *Journal of International Business Studies* 27, 753–779.

Niepce, W., and Molleman, E., 1996. Characteristics of Work Organization in Lean Production and Sociotechnical System. *International Journal of Operations and Production Management* 16, 77–90.

Niepce, W., and Molleman, E. 1998. Work Design Issues in Lean Production from a Sociotechnical Systems Perspective: Neo-Taylorism or the Next Step in Sociotechnical Design? *Human Relations* 51 (3), 259–287.

Norton, R., 1993. Ethno-Nationalism and the Constitutive Power of Cultural Politics: A Comparative Study of Sri Lanka and Fiji. *Journal of Asian and African Studies* 28 (3/4), 180–197.

Nunnally, J., and Bernstein, I. 1994. *Psychometric Theory*, 3rd ed. New York: McGraw-Hill.

Ohno, T., 1988. *Toyota Production System: Beyond Large-Scale Production*. Cambridge, MA: Productivity Press.

O'Reilly, C.A., and Chatman, J.A., 1996. Culture as Social Control: Corporations, Cults, and Commitment. In: B.M. Staw and L.L. Cummings, L.L. (Eds.), *Research in Organizational Behavior*, JAI Press, Greenwich, CT, pp. 157–200.

Papadopoulou, T.C., 2005. Leanness: Experiences from the Journey to Date. *Journal of Manufacturing Technology Management* 16, 784–807.

Pay, R., 2008. Everybody's Jumping on the Lean Bandwagon, but Many Are Being Taken for a Ride. *Industry Week* May.

Peetsma, T.T.D., 1993. Future Time Perspective as an Attitude: The Validation of a Concept. The 5th Conference of the European Association for Research on Learning and Instruction (EARLI), Aix-en-Provence, France.

Podsakoff, P.M., MacKenzie, S.B., Lee, J.Y., and Podsakoff, N.P., 2003. Common Method Biases in Behavioral Research: A Critical Review of the Literature and Recommended Remedies. *Journal of Applied Psychology* 88, 879–903.

Poulton, H., 2000. *Who Are the Macedonians?* Bloomington: Indiana University Press.

Power, D., Schoenherr, T., and Samson, D., 2010. The Cultural Characteristic of Individualism/Collectivism: A Comparative Study of Implications for Investment in Operations between Emerging Asian and Industrialized Western Countries. *Journal of Operations Management* 28, 206–222.

Power, D.J., and Sohal, A.S., 1997. An Examination of the Literature Relating to Issues Affecting the Human Variable in Just-in-Time Environments. *Technovation* 17, 649–666.

Pullman, M.E., Verma, R., and Goodale, J.C., 2001. Service Design and Operations Strategy Formulation in Multicultural Markets. *Journal of Operations Management* 19, 239–254.

Ralston, D.A., Holt, D.H., Terpstra, R.H., and KaiCheng, Y., 1997. The Impact of National Culture and Economic Ideology on Managerial Work Values: A Study of the United States, Russia, Japan, and China. *Journal of International Business Studies* 28, 177–207.

Raudenbush, S.W., and Bryk, A.S., 2002. *Hierarchical Linear Models,* 2nd ed. Thousand Oaks, CA: Sage.

Raudenbush, S.W., Bryk, A.S., Cheong, Y.F., Congdon, R., and du Toit, M., 2004. *HLM 6: Hierarchical Linear and Nonlinear Modeling.* Lincolnwood, IL: Scientific Software International.

Rinehart, J., Huxley, C., and Robertson, D., 1994. Worker Commitment and Labor-Management Relations under Lean Production at Cami. *Relations Industrielles-Industrial Relations* 49 (4), 750–775.

Robinson, A., and Schroeder, D., 2009.The Role of Front-Line Ideas in Lean Performance Improvement. *Quality Management Journal* 16 (4), 27.

Rosenberg, M.J., and Hovland, C.I., 1960. Cognitive, Affective, and Behavioral Components of Attitudes. In: C.I. Hovland and M.J. Rosenberg (Eds.), *Attitude Organization and Change*, Yale University Press, New Haven, CT, pp. 1–14.

Rother, M., 2009. *Toyota Kata: Managing People for Improvement, Adaptiveness and Superior Results.* New York: McGraw-Hill.

Rousseau, D., 1995. *Psychological Contracts in Organizations: Understanding Written and Unwritten Agreements.* Thousand Oaks, CA: Sage.

Rungtusanatham, M., 2001. Beyond Improved Quality: The Motivational Effects of Statistical Process Control. *Journal of Operations Management* 19 (6), 653–673.

Rungtusanatham, M., Anderson, J.C., and Dooley, K.J., 1997. Conceptualizing Organizational Implementation and Practice of Statistical Process Control. *Journal of Quality Management* 2 (1), 113–137.

Sakakibara, S., Flynn, B.B., Schroeder, R.G., and Morris, W.T., 1997. The Impact of Just-in-Time Manufacturing and Its Infrastructure on Manufacturing Performance. *Management Science* 43 (9), 1246*M*/1257.

Schein, E.H., 2004. *Organizational Culture and Leadership,* 3rd ed. San Francisco: Jossey-Bass.

Schonberger, R.J., 1982. *Japanese Manufacturing Techniques: Nine Hidden Lessons in Simplicity.* New York: Free Press.

Schonberger, R.J., 2007. Japanese Production Management: An Evolution—with Mixed Success. *Journal of Operations Management* 25, 403–419.

Schultz, K., Juran, D., Boudreau, J., McClain, J., and Thomas, L.J., 1998. Modeling and Worker Motivation in JIT Production Systems. *Management Science* 44, 1595–1607.

Schwartz, S., 1994. Beyond Individualism-Collectivism: New Cultural Dimensions of Values. In: U. Kim, H.C. Triandis, C. Kagitcibasi, S.-C. Choi, and G. Yoon (Eds.), *Individualism and Collectivism Theory, Method, and Applications,* Sage, Newbury Park, CA, pp. 85–122.

Seginer, R., and Schlesinger, R., 1998. Adolescents' Future Orientation in Time and Place: The Case of the Israeli Kibbutz. *International Journal of Behavioral Development* 22, 151–167.

Senge, P.M., 1991. *The Fifth Discipline: The Art and Practice of the Learning Organization.* New York: Doubleday.

Shah, R., and Ward, P.T., 2003. Lean Manufacturing: Context, Practice Bundles, and Performance. *Journal of Operations Management* 21 (2), 129–149.

Shah, R., and Ward, P.T., 2007. Defining and Developing Measures of Lean Production. *Journal of Operations Management* 25 (4), 785–805.

Shingo, S., 1985. *A Revolution in Manufacturing: The SMED System.* Cambridge, MA: Productivity Press.

Shook, J., 2009. Toyota's Secret. *MIT Sloan Management Review* 50, 30.

Shook, J., 2010. How to Change a Culture: Lessons from Nummi. *MIT Sloan Management Review* 51 (2), 63–68.

Siegel, J.I., and Larson, B.Z., 2009. Labor Market Institutions and Global Strategic Adaptation: Evidence from Lincoln Electric. *Management Science* 55 (9), 1527–1546.

Snijders, T., and Bosker, R., 1999. *Multilevel Analysis.* London: Sage Publications.

Spear, S.J., 1999. The Toyota Production System: An Example of Managing Complex Social/Technical Systems. 5 Rules for Designing, Operating, and Improving Activities, Activity-Connections, and Flow-Paths. Unpublished DBA, Harvard University, Cambridge, MA.

Spear, S.J., and Bowen, H.K., 1999. Decoding the DNA of the Toyota Production System. *Harvard Business Review* 77 (9 and 10), 97–106.

Sugimori, Y., Kusunoki, K., Cho, F., and Uchikawa, S., 1977. Toyota Production System and Kanban System Materialization of Just-in-Time and Respect-for-Human System. *International Journal of Production Research* 15 (6), 553–564.

Sullivan, L., 1896. *The Tall Office Building Artistically Considered.*

Sully de Luque, M., and Javidan, M., 2004. Uncertainty Avoidance. In: R.J. House, P.J. Hanges, M. Javidan, P.W. Dorfman, and V. Gupta (Eds.), *Leadership, Culture and Organization: The Globe Study of 62 Societies*, Sage, London.

Suzaki, K., 1985. Japanese Manufacturing Techniques: Their Importance to U.S. Manufacturers. *Journal of Business Strategy* (pre-1986) 5 (3), 10–19.

Suzaki, K., 1987. *The New Manufacturing Challenge: Techniques for Continuous Improvement.* New York: Free Press.

Tajiri, M., 1992. *TPM Implementation: A Japanese Approach.* New York: McGraw-Hill.

Tipuric, D., Podrug, N., and Hruska, D., 2007. Cultural Differences: Results from Empirical Research Conducted in Croatia, Slovenia, Bosnia and Herzegovina and Hungary. *Business Review, Cambridge* 7 (1), 151.

Treville, S.D., and Antonakis, J., 2006. Could Lean Production Job Design Be Intrinsically Motivating? Contextual, Configurational, and Levels-of-Analysis Issues. *Journal of Operations Management* 24 (2), 99–123.

Triandis, H.C., 1989. The Self and Social-Behavior in Differing Cultural Contexts. *Psychological Review* 96, 506–520.

Trommsdorff, G., 1983. Future Orientation and Socialization. *International Journal of Psychology* 18 (5), 381.

Trompenaars, F., and Hampden-Turner, C., 1998. *Riding the Waves of Culture: Understanding Cultural Diversity in Global Business*, 2nd ed. New York: McGraw-Hill.

Trovinger, S.C., and Bohn, R.E., 2005. Setup Time Reduction for Electronics Assembly: Combining Simple (Smed) and It-Based Methods. *Production and Operations Management* 14 (2), 205.

Tsuchiya, S., 1992. *Quality Maintenance: Zero Defects through Equipment Management.* Cambridge, MA: Productivity Press.

Tsui, A.S., Nifadkar, S.S., and Yi Ou, A., 2007. Cross-National, Cross-Cultural Organizational Behavior Research: Advances, Gaps, and Recommendations. *Journal of Management* 33, 426–478.

Turnbell, P., 1988. The Limits to Japanization-Just-in-Time, Labour Relations and the UK Automotive Industry. *New Tech, Work and Employment* 3 (1), 7–20.

Voss, C., and Blackmon, K., 1998. Differences in Manufacturing Strategy Decisions between Japanese and Western Manufacturing Plants: The Role of Strategic Time Orientation. *Journal of Operations Management* 16, 147–158.

White, R.E., Pearson, J.N., and Wilson, J.R., 1999. JIT Manufacturing: A Survey of Implementations in Small and Large U.S. Manufacturers. *Management Science* 45 (1), 1–15.

Whitten, P., and Collins, B., 1997. The Diffusion of Telemedicine—Communicating an Innovation. *Science Communication* 19, 21–40.

Whybark, D.C., 1997. GMRG Survey Research in Operations Management. *International Journal of Operations and Production Management* 17 (7–8), 686–696.

Whybark, C., Wacker, J., and Sheu, C., 2009. The Evolution of an International Academic Manufacturing Survey. *Decision Line* 40, 17–19.

Wiengarten, F., Fynes, B., Pagell, M., and de Búrca, S., 2011. Exploring the Impact of National Culture on Investments in Manufacturing Practices and Performance. *International Journal of Operations and Production Management* 31, 554–578.

Wincel, J. P., 2010. *Defying the Trend: Business Ethics and Corporate Morality from a Faith Perspective*. Holland, MI: God's Embrace Press.

Womack, J., Jones, D., and Roos, D., 1990. *The Machine that Changed the World*. New York: Rawson Associates.

Womack, J.P., and Jones, D.T., 1996. *Lean Thinking*. London: Simon and Schuster.

Womack, J.P., Jones, D., and Roos, D., 1991. *The Machine that Changed the World: The Story of Lean Production*, New York: Harper Perennial.

Wong, S., Bond, M.H., and Mosquera, P.M.R., 2008.The Influence of Cultural Value Orientations on Self-Reported Emotional Expression across Cultures. *Journal of Cross-Cultural Psychology* 39 (2), 224–229.

Yauch, C.A., 2000. *Moving towards Cellular Manufacturing: The Impact of Organizational Culture for Small Businesses*. Madison: University of Wisconsin–Madison.

Yauch, C.A., and Steudel, H.J., 2002. Cellular Manufacturing for Small Businesses: Key Cultural Factors that Impact the Conversion Process. *Journal of Operations Management* 20, 593–617.

Young, S.M., 1992. A Framework for Successful Adoption and Performance of Japanese Manufacturing Practices in the United States. *Academy of Management Review* 17, 677–700.

Zammuto, R.F., and O'Connor, E.J., 1992. Gaining Advanced Manufacturing Technologies' Benefits: The Roles of Organization Design and Culture. Academy of Management. *Academy of Management Review* 17, 701–728.

Glossary

Assertiveness (AS): The degree a society values confidence, confrontation, and aggressively defending one's position.

Future Orientation (FO): The degree a society sees present actions as affecting the future and long-term planning as useful for managing outcomes.

Gender Egalitarianism (GE): The degree a society seeks to minimize gender inequality.

Humane Orientation (HO): The degree a society encourages and rewards fairness, altruism, friendliness, generosity, caring, and kindness.

In-Group Collectivism (GC): The degree a society values loyalty, pride, and cohesiveness within a distinguishable group.

Lean Manufacturing (LM): A set of inter-related practices that focus on attaining continuous production flow (FLOW), having total preventative maintenance (TPM), allowing employee involvement (EMP), using pull production system (PULL), reducing setup times (SETUP), and utilizing statistical process control (SPC).

Lean Values: The underlying assumptions of lean manufacturing regarding the realities of production and what should be occurring.

National Culture: A society's shared set of fundamental assumptions, beliefs, and values regarding what exists and how people should behave.

Operating Performance (OP): Manufacturing performance as compare to competitors with respect to cost, quality, and delivery.

Performance Orientation (PO): The degree a society encourages members to excel among their peers.

Power Distance (PD): The degree a society expects organizational power to be concentrated at higher levels of an organization.

Uncertainty Avoidance (UA): The degree a society relies on procedures and norms to reduce the occurrence or likelihood of unpredictable events.

Index

M

O

P

Q

R

S

T

U

V

W

For Product Safety Concerns and Information please contact our EU representative GPSR@taylorandfrancis.com Taylor & Francis Verlag GmbH, Kaufingerstraße 24, 80331 München, Germany

Printed and bound by CPI Group (UK) Ltd, Croydon, CR0 4YY
01/05/2025
01858398-0001